THE LIGHT OF THE NIGHT SKY

GEOPHYSICS AND
ASTROPHYSICS MONOGRAPHS

AN INTERNATIONAL SERIES OF FUNDAMENTAL TEXTBOOKS

VOLUME 4

THE LIGHT OF THE NIGHT SKY

by

F. E. ROACH

Rutgers, The State University, Newark, N.J., U.S.A.;
University of Hawaii, Honolulu, H.I., U.S.A.

and

JANET L. GORDON

Bernice P. Bishop Museum, Honolulu, H.I., U.S.A.

D. REIDEL PUBLISHING COMPANY

DORDRECHT-HOLLAND / BOSTON-U.S.A.

First printing: December 1973

ISBN 978-90-481-8339-5

Published by D. Reidel Publishing Company,
P.O. Box 17, Dordrecht, Holland

Sold and distributed in the U.S.A., Canada, and Mexico
by D. Reidel Publishing Company, Inc.
306 Dartmouth Street, Boston,
Mass. 02116, U.S.A.

Printed in The Netherlands by D. Reidel, Dordrecht

TABLE OF CONTENTS

TABLES

PLATES

ILLUSTRATIONS

Zodiacal light seen from Haleakala, Hawaii, January 1967.

(Photograph by Courtesy of P. B. Hutchison, Dudley Observatory.)

PREFACE

Astronomy appears to us as a combination of art, science, and philosophy. Its study puts the universe into perspective, giving a sense of pleasure in its beauty, awe at its immensity, and humility at our trivial place in it.

From earliest human history, man has scrutinized the night sky – and wondered and marveled. With unaided eye but perceptive mind, he recognized order in the regular appearance and movements of individual objects, such as the planets and star groups (constellations), in their rhythmic and majestic progressions across the bowl of night. Even in the present era of scientific exactitude, there remains a profound awareness of mysteries beyond our present interpretations.

It is only in comparatively recent years, however, that man has recognized that it takes more than conventional astronomy to account for the beauties of the night sky. Radiations in the Earth's upper atmosphere provide a foreground light, the study of which has come under a new name, aeronomy. The science of aeronomy has rapidly burgeoned, and the student of the light of the night sky finds that he is involved in an interdisciplinary domain.

The 'meat' of one discipline, however, may be the 'poison' of the other. To the astronomer, the Earth's atmosphere, inhibiting his extra-terrestrial viewing, is a serious nuisance. To the aeronomer, the Moon, planets, stars, and Galaxies hamper his measurements and interfere with his studies of the Earth's upper atmosphere. Yet both sets of elements are basic to the beauties as well as to the understanding of the light of the night sky.

It is essentially the students of astronomy and aeronomy for whom we have written this book. We also hope, however, that it will present much of interest and value to the bemused sky watcher, for whom some detailed knowledge of the several con-
tried to meld these dual objectives to create a broadly based, professionally valid
tributors to the nighttime sky may increase his pleasure in contemplating it. We have
treatise that will lead the serious student to deeper probing into the phenomena and will inspire both him and the enthusiastic amateur to an appreciation of that half of their experience which we may refer to as their 'night life'.

As any scientific book goes to press, it becomes obsolete – knowledge continues to expand along with the expanding universe. Superimposed on the obsolescence inherent in an active subject being reported at a point in time is another; the night sky cannot be considered constant in time scales – astronomical, geological, or even historical. Our view toward any part of the Galaxy changes for two reasons – first, the differential rotation of layers outward from the center and, second, the fact that the intrinsically very bright stars have lifetimes relatively short, measured in only millions of years. Thus during a period of one galactic rotation of 200000000 yr (at our distance from the galactic center) – less than 100 times man's existence on Earth – the appearance of the Milky Way must have changed dramatically.

In Chapter 6 we call attention to the fact that the zodiacal cloud, which is responsible for an important component of the light of the night sky, has a lifetime in the order of 80000 years. Astronomically speaking, this is a short-lived phenomenon.

A factor somewhat under human control is the radiation from the upper atmosphere. Already it has been measurably affected several times by changes in the ionization of the upper atmosphere from high-altitude nuclear detonations. Lower-atmosphere contaminations due to the 'civilization' crawling over the planet's surface could conceivably affect our viewing of the night sky in the centuries ahead.

Our description of the several components contributing to the light of the night sky applies to the present time and to our particular cosmic location. We thus present a study of an ephemeral and parochial phenomenon.

Honolulu, August 1973

CHAPTER 1

FROM DAY TO TWILIGHT TO NIGHT

Life on planet Earth is tuned to the diurnal drama of day-twilight-night-dawn-day. As Earth rotates within the sunlight sheath, the outward-looking observer notes changes of brightness which cover many powers of ten. The day sky, in the absence of clouds, appears as a blue vault resulting from the scattering of sunlight by the atmospheric components. The brightness of the sky decreases rapidly after sunset as the solar rays illuminate only the higher regions of the atmosphere above the observer. At a solar depression of 18° (end of astronomical twilight) the amount of illuminated atmosphere has become so small that the scattered sunlight is trivial and the night sky can be observed without twilight competition (Figure 1-1).

1.1. The Day Sky – The Blue Planet

The blueness of the day sky was explained by Lord Rayleigh III in his classical paper, 'On the Light from the Sky, Its Polarisation and Colour' (1871), the key portion of which is presented in Plate I. Analytically, he found that, for small particles,

$$\frac{I}{I_0}=\frac{\text{const.}\,(1+\cos^2\theta)}{\lambda^4}. \tag{1.1}$$

Since the human eye is sensitive over the octave from about 7000 Å (7×10^{-5} cm) to 3500 Å (3.5×10^{-5} cm), the ratio of intensity of scattered light for the shorter (bluer) wavelength relative to the longer (redder) wavelength is

$$\frac{I_{\text{blue}}}{I_{\text{red}}}=\left(\frac{\lambda_{\text{red}}}{\lambda_{\text{blue}}}\right)^4=\left(\frac{7000}{3500}\right)^4=\left(\frac{2}{1}\right)^4=16. \tag{1.2}$$

In Equation (1.1) the term $(1+\cos^2\theta)$ is due to two conditions of polarization – the 1 corresponds to the electric vector perpendicular to the plane of scattering, and the $\cos^2\theta$ to the electric vector parallel. The situation is best shown in the form of a polar diagram (Figure 1-2). Most of the incident radiation will pass the particle in its path, but a small fraction will be scattered. The relative intensity of the total scattered light as a function of scattering angle is shown as the solid line $(1+\cos^2\theta)$. The two polarized components are shown as dashed lines. Polar diagrams of scattered light from particles comparable in size to the wavelength (such as dust) will be discussed later in connection with two astronomical phenomena – the zodiacal light and the diffuse galactic light.

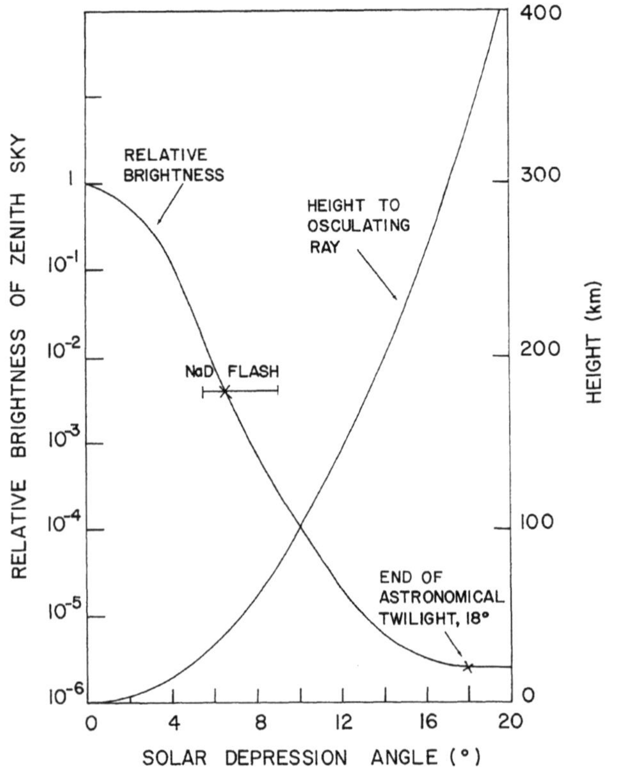

Fig. 1-1. Relative brightness of zenith sky during twilight (5200 Å; after Ashburn, 1952:85), and height to lowest part of zenith atmosphere illuminated by Sun during twilight.

1.2. The Twilight Sky

As Earth's rotation carries the Sun from the horizon to an 18° depression angle, the zenith brightness decreases about 400000-fold, as shown in Figure 1-1. In the tropics, this darkening can occur in as short a time as 72 min.

With the increasing depression angle, progressively higher levels of the upper atmosphere are being irradiated by the Sun, until the night level of 'brightness' is achieved. The general spectroscopic nature of the light during the twilight phase is that of scattered sunlight. A color photograph during midtwilight reveals a blue sky long after the eye has lost its color sensitivity, showing that the mechanism is still Rayleigh scattering.

At a solar depression of about 6° the lower boundary of the solar radiation is at 35-km altitude. Solar radiation impinging on the sodium atoms above this region excites them to the $^2P^0$ state, which results in the emission of the yellow Na D lines. At a solar depression of 6.5°, the general twilight radiation has been reduced to such a low

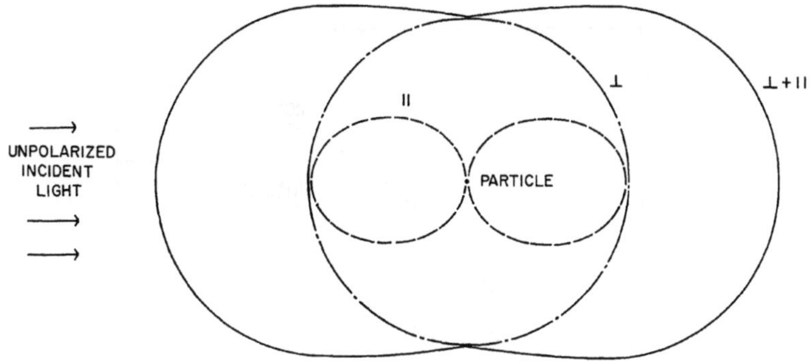

Fig. 1-2. Polar diagram of scattering of unpolarized incident light by small particle (e.g., molecule). The components with electric vector perpendicular (⊥) and parallel (||) to the scattering plane, and their sum are shown.

photometric level that the emission lines are of intensity equal to that of the absorption Na D lines in the solar spectrum (Figure 1-1). The lines fade in brightness rapidly as the increasing solar-depression angle puts the osculating ray above the level at which the Na atoms are sufficiently numerous to produce an observable line. The relatively short duration of the phenomenon has led to its description as the 'sodium flash'. A similar flash occurs during twilight for the red (6300 Å, 6364 Å) lines of O.

1.3. The Human Eye and Its Adaptation to Low Light Levels

The eye, with its remarkable ability to adapt to the very wide range of brightness from the day to the night sky levels, adds a dimension of aesthetic enjoyment to the 'sky watcher'.

The dark adaptation of the eye is a complicated process involving a small factor (of about seven) that is due to the opening up of the iris to admit more light, and a very large factor (which can be more than 10^3, for a large field) that results from a chemical change in one of the two types of light receptors in the retina. These receptors – named rods and cones because of their shapes – serve different functions.

The approximately 7×10^6 cones are concentrated in the central part of the retina, directly behind the lens, and provide very sharp visual definition and color sensitivity over a small field – 2 to 5° – at high light levels. They decrease in frequency outward from that center.

The 13×10^7 rods have the complementary distribution, being sparse near the center of the retina but increasing in numbers and concentrating as they approach the periphery of the retina. They generate the unique chemical 'visual purple', which they accumulate in the absence of light and which seems to be the active agent in achieving dark-adapted vision over a period of time. The rods detect faint light and movement more effectively than do cones. Thus, in dark adaptation, they are the active element.

Figure 1-3 shows the relative sensitivity of the eye to dark adaptation as a function of time for five centrally fixated fields varying from 2 to 20°. It is striking to realize that,

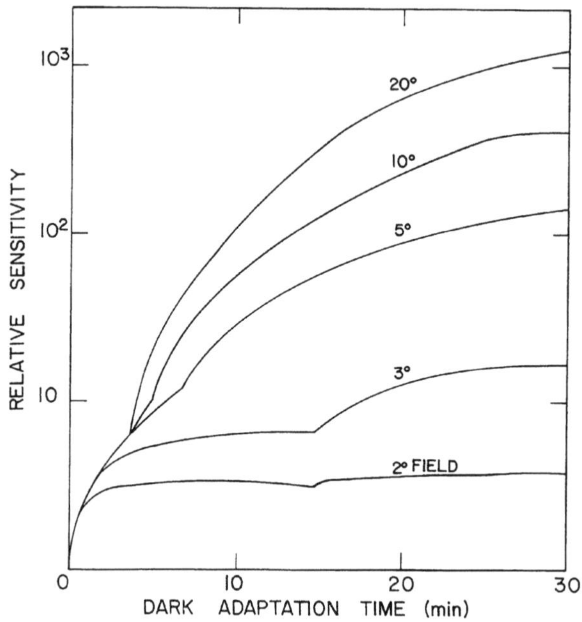

Fig. 1-3. Increase in visual sensitivity with dark-adaptation time for a centrally fixated field of variable diameter.

over the wider fields – from 5 to 20° – the relative sensitivity increases between 10^2 and 10^3 times in a period of 30 min. Over the smaller fields, in which the cones are concentrated, the sensitivity increases only slightly more than 10 times, and this increase is probably primarily due to the effect of the opening of the iris.

A familiar example is the experience of entering a dark movie theater from brightly sunlit outdoors – you seem *blind*! After groping to a seat, however, you notice that you can begin to detect, dimly, your surroundings. And after a few more minutes you can clearly observe details – you are dark-adapted.

A particular application of the problem of detection of discrete light sources such as stars in competition with the background of the twilight sky is illustrated in Figure 1-4, in which it is shown that fainter and fainter stars can be detected as the background light becomes dimmer. Under excellent sky conditions, stars can be detected to a level between visual magnitudes 5 and 6 at the end of twilight.

The detection of a point source such as a star and of an extended source covering many degrees, such as the gegenschein (to be discussed in Chapter 3) emphasizes the different aspects of visual perception. As described above, the cones are concentrated in the center of the retina and each cone is connected to the brain by a separate nerve; hence vision is very distinct when the light is bright so that the iris closes down and the center of the retina is used. The rods' chemical change during dark adaptation, and their concentration toward the outer part of the retina, permit improved detection of

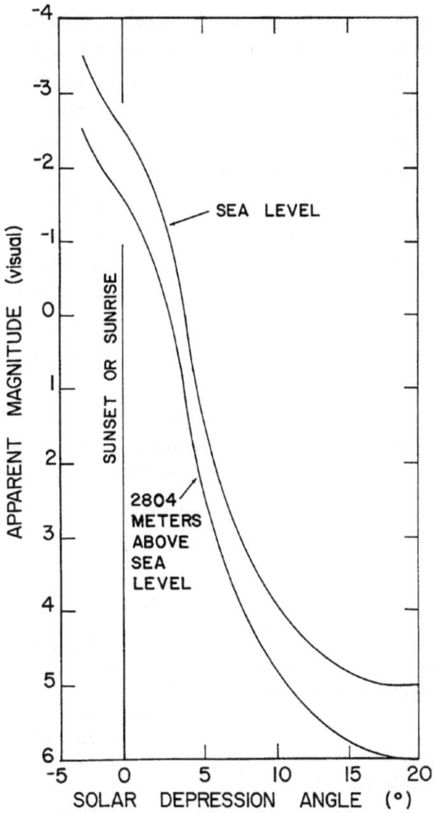

Fig. 1-4. Detectability of stars and planets during twilight. Between sunset and a solar depression angle of 7° the threshold at sea level changes from $m = -2.4$ to $+2.6$, a change of 10^2 in brightness.

faint extended sources by (1) a dark-adaptation period of up to 30 min and (2) preferential use of the outer parts of the retina (averted vision). The detection of stars during twilight is a problem of contrast plus dark adaptation.

1.4. The Night Sky

Everything considered, Earth dwellers are extraordinarily fortunate in that the several separate and distinct components of the light of the night sky can be disentangled, at least to a reasonable extent. We shall explore later some unhappy circumstances that might have made such an analysis of resolving the components of the night-sky light difficult or even impossible.

The several components of the light of the night sky are listed in Table 1-I, together with their physical origins and their coordinate systems. Note that three different coordinate systems are applicable to the several components. This situation is analogous to a case of physical measurement in which three variables are involved, each of which is a function of one of the three coordinates in a Cartesian system. In our case, we have complications in that two of the components, the airglow and the aurora, have strong time dependence. The viewer can effectively eliminate the aurora, however, by avoiding the vicinity of the auroral zone.

A first, gross separation of the components by virtue of their dependence on different coordinate systems is illustrated in Figure 1-5. Several thousands of zenith observations at Haleakala (Island of Maui, Hawaii) have been assembled. The temporal airglow changes have been averaged; the zodiacal light has measurable variations as the ecliptic latitude changes during the year; and the integrated starlight shows two peaks of brightness as the two arms of the Milky Way cross the Haleakala zenith at sidereal times 90° (6 hr) and 295° (19 hr 40 min). In this plot we do not distinguish the minor constituents – the diffuse galactic light and the integrated cosmic light – since their detection calls for a very detailed attack on the data. In Figure 1-6 we show a graphical summary of the distribution of the components of the light of the night sky at a location well away from the auroral zone.

TABLE 1-I

Components of the light of the night sky

Component	Function of		Physical origin
	Position in sky	Other	
Integrated starlight	b, l;[a] galactic coordinates		Stars of our Galaxy
Zodiacal light	$\beta, \lambda - \lambda_\odot$; ecliptic coordinates		Sunlight scattered by interplanetary dust
Airglow	Z, A; horizon coordinates	Local time Latitude Season Solar activity Wavelength Height	Ambient excitation (usually photochemical) of upper-atmosphere atoms and molecules
Aurora		Magnetic latitude Season Solar activity Magnetic activity Wavelength	Excitation of upper-atmosphere atoms and molecules by energetic particles
Diffuse galactic light	b, l		Scattering of starlight by interstellar dust
Integrated cosmic light	b, l	Cosmological red shift Cosmological model	

[a] b = galactic latitude; l = galactic longitude; β = ecliptic latitude; λ = ecliptic longitude; $\lambda_\odot$ = ecliptic longitude of Sun; Z = zenith angle; A = azimuth.

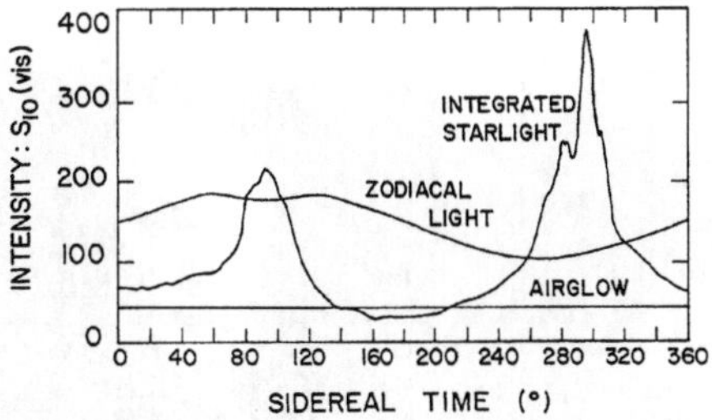

Fig. 1-5. Resolution as a function of sideral time of total zenith intensities at Haleakala (Hawaii) into three components – airglow, zodiacal light, and integrated starlight. In the latitude of Haleakala (about 20°), the ecliptic is in the zenith twice during the year – at 60° (4 hr) and 120° (8 hr) – and thus the zodiacal light shows two slight maxima.

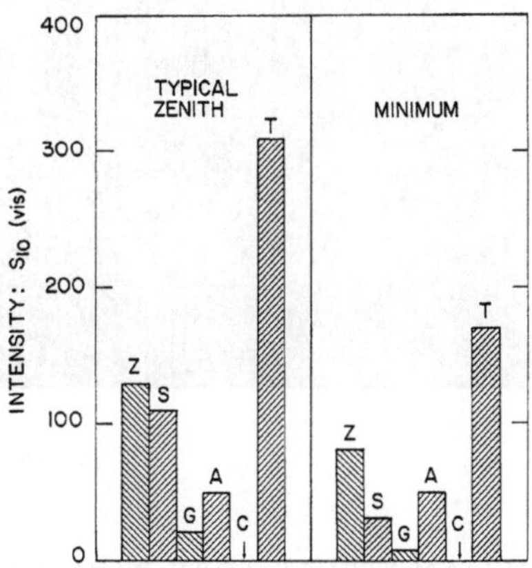

Fig. 1-6. Components of the light of the night sky. Minimum values for zodiacal light, starlight, and the airglow occur at the ecliptic pole, the galactic pole, and the zenith, respectively. Z = zodiacal light; S = integrated starlight; G = diffuse galactic light; A = airglow; C = cosmic light; T = total.

The plan of the book is to discuss the five components of the light of the night sky. We shall concentrate on the visual part of the spectrum, giving brief mention, however, to recent satellite observations in the near-infrared and ultraviolet. In addition to describing the observational facts, we shall discuss some of their implications and interpretations.

References

Ashburn, E. V.: 1952, *J. Geophys. Res.* **57**, 85.
Rayleigh III, John William Strutt, Baron: 1870 (Ms.), 16.
Rayleigh III, John William Strutt, Baron: 1871, *Phil. Mag.*, February.
Rayleigh III, John William Strutt, Baron: 1877, *Theory of Sound* (Macmillan, London).
Rayleigh IV, Robert John Strutt, Baron: 1968 (augmented edition), *Life of John William Strutt, Third Baron Rayleigh, O.M., F.R.S.* (University of Wisconsin Press, Madison).

Lord Rayleigh III
(Courtesy of the Royal Society)

PLATE I JOHN WILLIAM STRUTT, LORD RAYLEIGH III
1842–1919

Lord Rayleigh III is known as 'the scattering Rayleigh', to distinguish him from his son, Lord Rayleigh IV, who is dubbed 'the airglow Rayleigh'. Rayleigh III, however, distinguished himself across a broad spectrum of the science of a century ago. He might also be called 'the sound Rayleigh', as his *Theory of Sound*, published in 1877, is still one of the most respected books on the subject.

Born on November 12, 1842, John William Strutt was descended from a distinguished and well-to-do British family. His grandfather, Colonel John Holden Strutt, had been granted the peerage at the coronation of King George IV as a reward for outstanding public service. He had requested that the title be awarded in the person of his wife, Lady Charlotte Strutt. The name Rayleigh was apparently somewhat arbitrarily chosen, although Colonel Strutt did own some property near the small town of Rayleigh in Essex. His son John James became Lord Rayleigh II on the death of Lady Charlotte in 1836. No scientific interest or talent seems to have been apparent in the family to this time.

The childhood of Rayleigh III was characterized by poor health and lively curiosity. Not until his undergraduate years at Cambridge, however, did his genius for mathematics and science appear. A gifted theoretician, he verified his mathematical results with painstaking experimental work, and one of his greatest laments, both at Cambridge and later in his own home, was for the lack of laboratory equipment. He finally hired a man to constuct items to his specifications for his home laboratory.

More than 100 yr ago, in 1871, Rayleigh III published his classical paper on scattering, 'On the Light from the Sky – Its Polarisation and Colour' (1871), in which he explained the blueness of the day sky. The law governing the effect, stated below, is reproduced from his original handwritten draft (1870, Ms.)

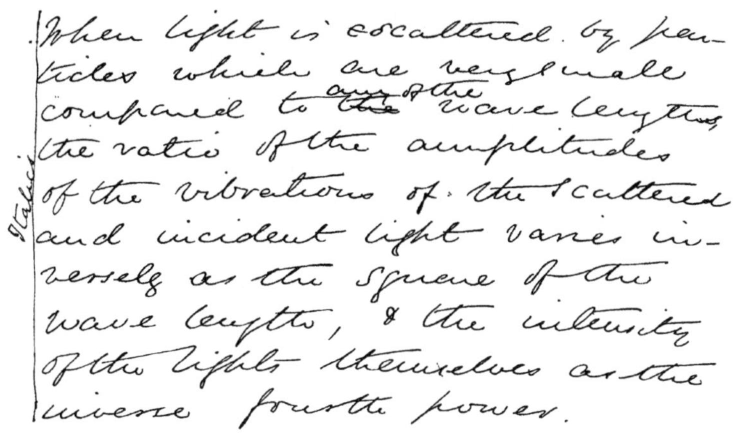
Italics

When light is scattered by particles which are very small compared to ~~the~~ any of the wave lengths the ratio of the amplitudes of the vibrations of the scattered and incident light varies inversely as the square of the wave lengths, & the intensity of the lights themselves as the inverse fourth power.

From 1879 through 1884, Rayleigh held a Cambridge professorship, but then returned to his home in Terling to conduct his experiments and develop his theories. During this period he published a paper in *Nature* on the subject of mechanical flight, and in 1900 he lectured on 'The Mechanical Principles of Flight'.

In 1887, Rayleigh accepted the Professorship of Natural Philosophy at the Royal Institution, where he presented lectures illustrated with experiments of his own devising – on such varied subjects as sound and vibration, optics, electricity and magnetism, light and color, the mechanical properties of bodies, matter at rest and in motion, the forces of cohesion, and polarized light. He resigned the Professorship in 1905 but gave occasional lectures after that time.

Throughout his life he was not only friend and co-worker with the major scientists of his period – A. J. Ångstrom, Robert Boyle, Niels Bohr, Charles Darwin, Albert Einstein, Michael Faraday, Willard Gibbs, Hermann Helmholtz, Sir John Herschel, Lord Kelvin, Lister, Clerk Maxwell, A. A. Michelson, Sir William Ramsey, Ernest Rutherford, Sir Frederick Stokes, John Tyndall, and Sir J. J. Thomson – but also with many high-level social, intellectual, and political figures, such as Winston S. Churchill, Joseph Chamberlain, Arthur Balfour, and Lord Salisbury.

His later years were characterized by a broadening of his activities into more public work – local political administration, matters of public concern – but he remained devoted to science to the end of his life, in 1919.

Rayleigh's old friend Sir J. J. Thomson presented the memorial address in Westminster Abbey; a few excerpts from his address may summarize Rayleigh III most effectively:

> "Among the 446 papers ... there is not one that is trivial ... and among that great number there are scarcely any which time has shown to require correction .... The first impression we gain ... is the catholicity of Lord Rayleigh's work .... Lord Rayleigh took physics for his province and extended the boundary of every department of physics ... in the majority of his writings we have a combination of mathematical analysis and experimental work, and his papers ... afford the best illustration of the true co-ordination of these two great branches of attack on the problems of nature. The physical ideas direct the mathematical analysis into the shortest and most appropriate channels, while the mathematics gives precision and point to the physics." (Rayleigh IV, 1968, pp. 309–310.)

Having made unique scientific contributions through about 55 of his 77 years, Lord Rayleigh III merited the Memorial in Westminster Abbey, on which the inscription reads:

"JOHN WILLIAM STRUTT, O.M., P.C.

3rd Baron Rayleigh

Chancellor of the University of Cambridge, 1908–1919

President of the Royal Society, 1905–1908

An Unerring Leader in the Advancement of Natural Knowledge."

(Rayleigh IV, 1968: 374)

CHAPTER 2

STAR COUNTS AND STARLIGHT

In Chapter 1 we found that the unaided eye could distinguish stars to the 6th visual magnitude from a mountain top and to the 5th magnitude from sea level. Over the entire sky there are 4850 stars brighter than magnitude 6 and 1620 brighter than magnitude 5. These are manageable numbers; but with telescopic aid the number of stars either visible or photographable increases enormously. For example, there are

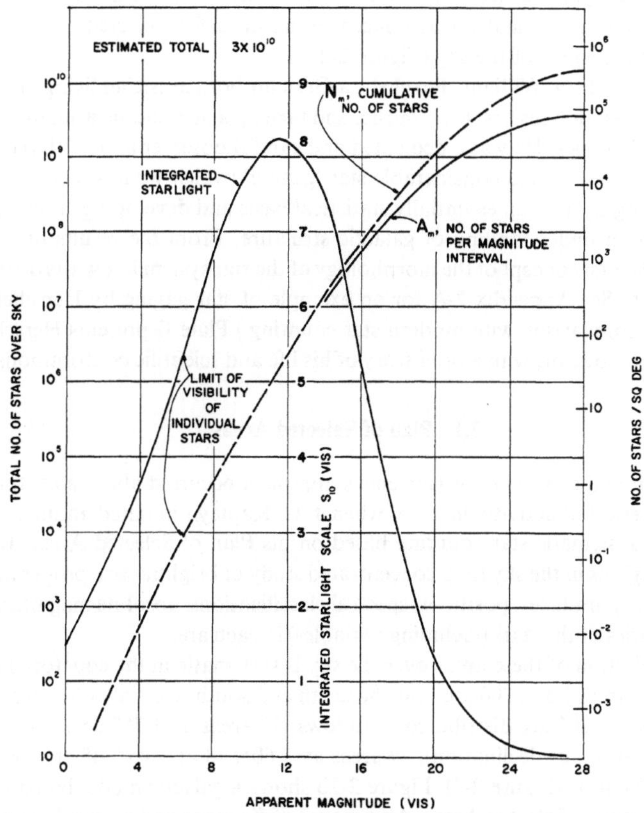

Fig. 2-1. Plots of: (a) number of stars per magnitude interval (A_m), (b) cumulative number of stars (N_m), and (c) integrated starlight (J_m) per magnitude interval as a function of apparent visual magnitude. (Based on MW301, averaged over the whole sky.)

10^6 stars brighter than visual magnitude 11.2 and 10^9 brighter than magnitude 20. A definitive count is possible for the brighter stars, but for the fainter ones counting has been completed only for samples.

The entire sky has been photographed using telescopes designed to include large regions in good focus. The National Geographic Society-Palomar Observatory Sky Survey consists of 1036 separate reproductions from original plates made with the Palomar 48-in. Schmidt telescope for that part of the sky photographable from the Palomar latitude. Stars as faint as photographic magnitude 20 (visual magnitude 18.8) appear on the reproductions. Figure 2-1 indicates that there are of the order of 5×10^8 stars brighter than visual magnitude 18.8. Note, however, that this is less than 2% of the total number of stars in our Galaxy.

The task of counting and tabulating this vast reservoir of information is formidable. Historically, it has been approached in two ways – (1) by starting with the less-numerous bright stars, and (2) by concentrating on a few selected, and hopefully representative, regions of the sky (Figure 2-2).

In the 18th century, William Herschel – a German-born musician living in England – decided to abandon his first love, music, and take up a new enthusiasm, the building and use of telescopes. He was a dedicated and careful observer and, with the help of his sister Carolina, he did considerable star 'gaging' (which we now call star counting), compiling data on an essentially statistical basis and developing an imaginative approach to an understanding of galactic structure. From the results of his early work grew the first concept of the morphology of the star system known as our Galaxy (Figure 2-3a). (See Appendix 2-A for an example of star gaging by Herschel and a quantitative comparison with modern star counting.) Plate II presents Herschel, the outstanding innovator, with a brief story of his life and scientific contributions.

2.1. Plan of Selected Areas

An extension and expansion of Herschel's approach occurred about a century after Herschel's period of activity, in 1906, when J. C. Kapteyn initiated an international program of systematic star counting based on his Plan of Selected Areas. Kapteyn chose 206 regions in the sky for a concentrated study of brightnesses, proper motions, trigonometric parallaxes, positions, spectral classifications, absolute magnitudes, and radial velocities of the stars (including variables) in each area.

The distribution of these areas over the sky is systematic in the equatorial coordinate system. Areas 1 and 206 are near the north and south celestial poles respectively. Areas 2 through 205 are distributed as follows: 12 areas at $\pm 75°$ declination (six at plus and six at minus declination), 24 areas at $\pm 60°$, 48 at $\pm 45°$, 48 at $\pm 30°$, 48 at $\pm 15°$, and 24 at 0° (Figure 2-4). Figure 2-3b shows a galactic model based on early work on Kapteyn's Selected Areas. An account of Kapteyn and his work is presented in Plate III.

Through the 20th century, since he initiated the Selected Areas concept in 1906, productive activity on Kapteyn's method has continued steadily.

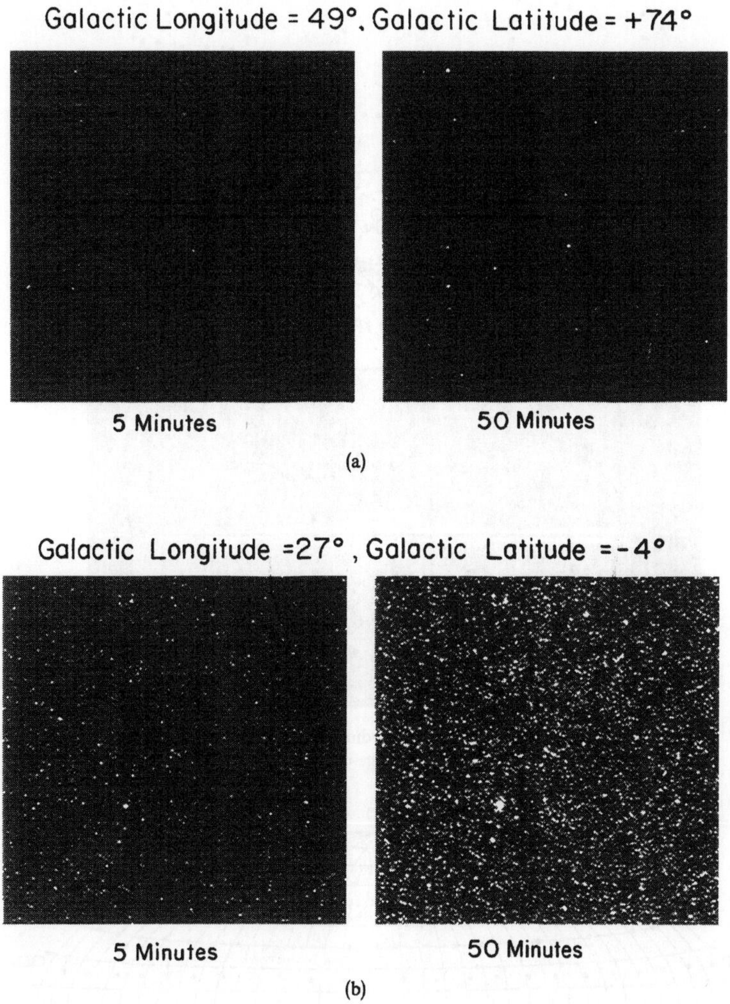

Fig. 2-2. Photographs of star-poor (2-2a) and star-rich (2-2b) fields, at short (5 min) and long (50 min) exposure times. The 2-2a field is about 16° from the North galactic pole ($\alpha = 14$ h 01 m; $\delta = +30^{\circ}.5$) and is Selected Area 58, labelled A on Figure 2-4. The star-rich field is 4° from the galactic equator, in the constellation Scorpius ($\alpha = 18$ h 55 m; $\delta = -7^{\circ}.0$). The field size of each photograph is 16′ by 16′, or 0.07111 sq deg. The effect on the number of stars visible as the exposure time is increased is evident by comparing the photographs left and right. Counts of the number of stars on the working prints gave 16 for a(5), 44 for a(50), 930 for b(5), and 4500 for b(50). Referring these counts to the quantity N_m (number of stars brighter than magnitude m per sq deg – see p. 17), we get $\log N = 2.4$ for a(5), 2.8 for a(50), 4.1 for b(5), and 4.8 for b(50). The photographic limiting magnitude is about 16 for the 5 min exposures and about 18 for the 50 min exposures. The photographs were made by B. J. Bok with the $f/9$ Cassegrain arrangement of the Steward Observatory 90 in. reflector at Kitt Peak.

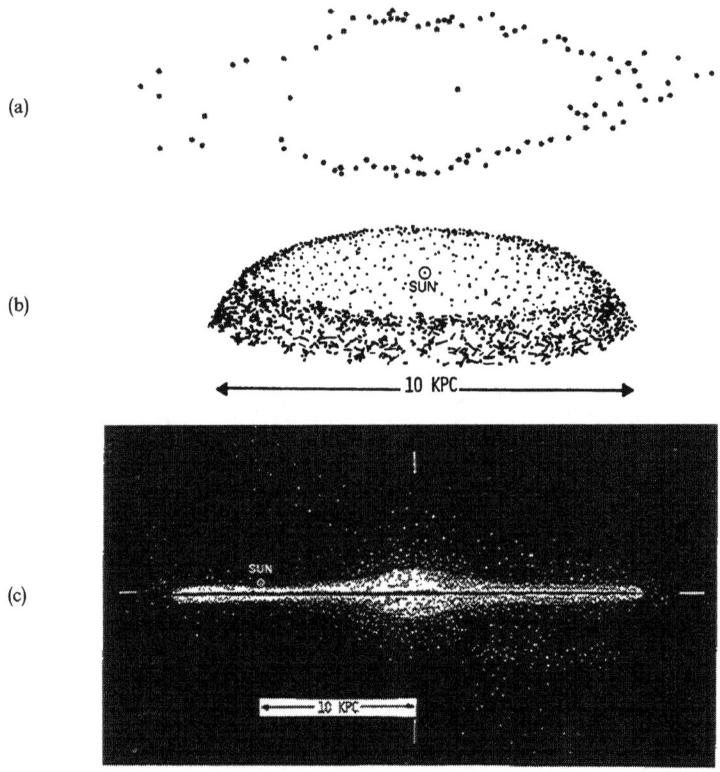

Fig. 2-3. Developing concepts of our Galaxy according to: (a) Herschel, (b) Kapteyn, and (c) present (seen edge-on).

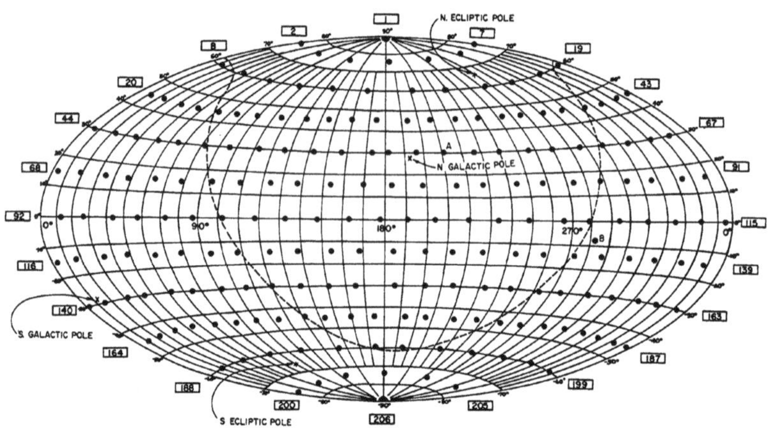

Fig. 2-4. Distribution of Kapteyn's Selected Areas, shown on equatorial coordinates; the galactic equator is shown as the dashed line. Region A (near the galactic pole) is the star-poor field of Figure 2-2a; region B (near the galactic equator) is the star-rich field of Figure 2-2b.

2.2. Modified Star Counting

An understanding of the morphology of our Galaxy has been significantly advanced during the last two decades by a continuation of the star-counting technique as a probe. The counting approach has departed from the gross statistical procedures characteristic of the earlier work and has concentrated on detailed studies in which a depth inference in the line of sight is made from the observed distribution of the brightnesses of stars projected against the celestial sphere, together with measurements of their spectral characteristics*.

The analysis is complicated by the fact that the apparent brightness of a star of

Fig. 2-5. Spiral Galaxy Messier 51. Our Galaxy would probably look similar to M51 if seen face-on from a comparable distance. The approximate location of our Sun would be about two-thirds of the way out from the center, in one of the spiral arms (see Figure 2-6).

* Herschel was handicapped in his early attempts to obtain even an approximation to a depth analysis because, in the absence of any knowledge as to the very wide distribution of intrinsic stellar brightnesses as revealed by spectroscopic means, he used the gross oversimplification that 'brightness equals nearness'.

given intrinsic luminosity depends on both its distance and its partial obscuration by dust between star and observer. (An assist has come from the radio astronomers, who have taken advantage of the fact that radio waves penetrate through the ubiquitous dust in our Galaxy much more efficiently than do the optical radiations.)

An example of such an analysis is the present picture of our Galaxy as a spiral similar to the many external spirals that have been photographed by our major telescopes (Figure 2-5). A somewhat schematic illustration of three spiral arms that have been identified in the part of our Galaxy near our Sun is shown in Figure 2-6. The Sun

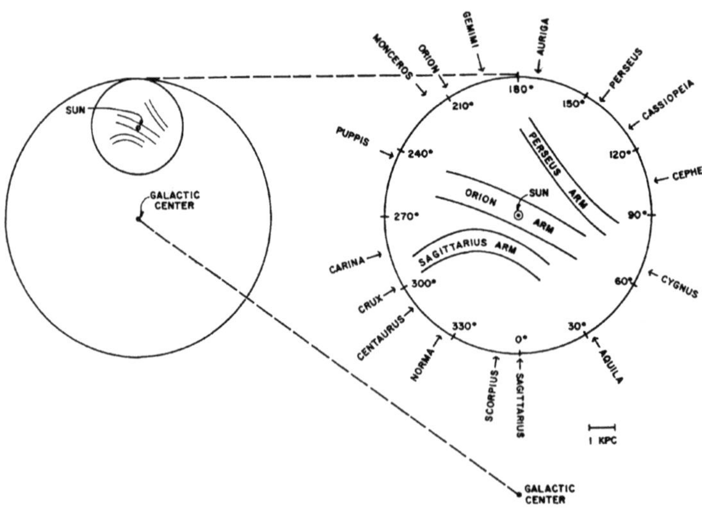

Fig. 2-6. Spiral structure of the portion of the Galaxy near the Sun, based on optical and statistical studies. The page is in the galactic plane. Galactic longitudes, l^{II}, are indicated around the circumference of a circle centered on the Sun, together with some of the constellations visible to a solar-terrestrial observer. The direction toward the galactic center (10 kpc distant) is at $l^{II}=0°$ in the constellation of Sagittarius.

is embedded in one of the spiral arms – the so-called Orion arm. The Sagittarius arm and the Perseus arm are as near as 1 and 2 kpc (kiloparsecs), respectively, and have been delineated out to several kpc. For reference, the distance of our Sun from the center around which the Galaxy revolves is about 10 kpc.

The use of such a probing of an arm in our spiral-type Galaxy has led to major contributions to our knowledge in recent years by Bart J. Bok, who is still exploring this productive technique (Plate IV).

2.3. Distribution of Integrated Starlight over the Sky

In Table 2-I and Figure 2-1, we show some statistical results based on a study of Selected Areas 1 to 139 (Seares *et al.*, 1925). The data are in visual magnitudes and refer to the mean over the sky. The Selected Areas chosen included the zone as far

TABLE 2-I

Summary of star counts by visual-magnitude intervals; mean over the sky (from MW301[a])

m(vis)	$\log A_m$	$\Delta \log A_m$	J_m (S_{10}(vis))	J_m(cum) (S_{10}(vis))	$\log N_m$	$\log T_m$
0	−3.821		1.51	105.20		
1	−3.301	0.520	1.99	103.69		
2	−2.807	0.494	2.47	101.70		
3	−2.326	0.481	2.98	99.23		
4	−1.841	0.485	3.62	96.25	−1.888	2.727
5	−1.362	0.479	4.34	92.63	−1.405	3.210
6	−0.897	0.465	5.05	88.29	−0.929	3.686
7	−0.431	0.466	5.88	83.24	−0.461	4.154
8	+0.023	0.454	6.65	77.36	+0.001	4.616
9	0.466	0.443	7.34	70.71	0.453	5.068
10	0.897	0.431	7.90	63.37	0.895	5.510
11	1.312	0.415	8.17	55.47	1.324	5.939
12	1.714	0.402	8.20	47.30	1.740	6.355
13	2.097	0.383	7.89	39.10	2.141	6.756
14	2.461	0.364	7.26	31.21	2.525	7.140
15	2.805	0.344	6.38	23.95	2.891	7.506
16	3.120	0.315	5.25	17.57	3.235	7.850
17	3.411	0.291	4.08	12.32	3.557	8.172
18	3.677	0.266	3.00	8.24	3.856	8.471
19	3.916	0.239	2.07	5.24	4.132	8.747
20	4.131	0.215	1.35	3.17	4.385	9.000
21	4.318	0.187	0.83	1.82		
22	4.481	0.163	0.48	0.99		
23	4.618	0.137	0.26	0.51		
24	4.729	0.111	0.14	0.25		
25	4.815	0.086	0.06	0.11		
26	4.875	0.060	0.03	0.05		
27	4.910	0.035	0.01	0.02		
28	4.919	0.009	0.01	0.01		(10.48)

[a] Entries corresponding to m(vis) > 20 are based on a quadratic extrapolation of the data in MW301 (Seares *et al.*, 1925).

south as declination −15°, which can be observed at Mt. Wilson during meridian passage at a zenith angle of 49°. (The next zone to the south, at a declination of −30°, has a zenith angle of 64° for meridian passage; although observable, this angle is a little low in the sky for accurate photometry.) Much of the bright Milky Way in the southern sky is not included in the Mt. Wilson study, but the over-all results can be profitably examined.

In Figure 2-1 we have used three scales for the ordinate. To the left is indicated the total number of stars over the sky; to the right, the number per square degree. Both scales can be referred to either the cumulative number of stars, N_m (number of stars brighter than magnitude m), or to the number per magnitude interval, A_m (number of stars between magnitudes $m-\frac{1}{2}$ and $m+\frac{1}{2}$). The central scale applies to the integrated starlight, J_m, the brightness in S_{10}(vis) units from the stars between magnitudes $m-\frac{1}{2}$ and $m+\frac{1}{2}$. The relationship between A_m and J_m is

$$J_m = A_m \cdot 10^{-0.4(m-10)}.$$

The plot of J_m is of direct interest to the topic of this book, since the summation under the curve gives the mean contribution of starlight to the light of the night sky. J_m goes through a maximum at m(vis)=12. The position of the maximum depends on the point of inflection in the $\log A_m$ vs m curve, where $(\Delta \log A_m)/\Delta m = 0.4$. When $(\Delta \log A_m)/\Delta m$ is greater than 0.4, the number density of stars increases more rapidly than their brightness decreases, and J_m increases with m; when $(\Delta \log A_m)/\Delta m$ is less than 0.4, J_m decreases with m. The maximum occurs for $(\Delta \log A_m)/\Delta m = 0.4$.

The summation of J_m (column 5 in Table 2-I) is made from the faint end, since we are interested in evaluating the background starlight with respect to observers for whom stars of different brightnesses (magnitudes) are discretely observed and hence do not contribute to the 'interstellar' background. For example, the observer who scans the sky without a telescope will see discrete stars to the 5th or 6th mag., and the background starlight *on the average* is about 90 S_{10}(vis) units. An observer using a 4-in. telescope sees stars as faint as magnitude 12, and the background of unresolved stars will be only some 47 S_{10}(vis). With a 100-in. telescope, stars of magnitude 19 are discernible, and the background starlight is reduced to about 5 S_{10}(vis).

2.4. Distribution of Starlight over the Sky

In the preceding paragraphs we have considered the average background of light from unresolved stars. The departure from the average is of interest to the investigator of galactic structure but also to us, because we are concerned with the distribution of starlight over the sky.

As mentioned earlier, the fact that the galactic system concentrates the stars toward the galactic equator (Milky Way) facilitates the disentanglement of integrated starlight from the other major constituents of the night-sky light. We plan to pursue this matter with reference to two compilations of star counts – (1) the Mt. Wilson publication 301 (Seares *et al.*, 1925) previously mentioned, and (2) Groningen Astronomical Observatory publication 43 (van Rhijn, 1925). The former is based on Selected Areas 1 through 139, the latter on the totality of the Areas.

In Figure 2-7 we show plots of J_m for galactic latitudes $b=0°$ and $80°$, from the two published sources. The data are assembled in Table 2-II. Figure 2-8 displays the relationship between J_m(cum) and galactic latitude for the mean of GR-43 and MW-301 (from columns 4 and 8 of Table 2-III). In general, the integrated starlight is about nine times as bright in the Milky Way as near the galactic pole. The fact that the Milky Way does not appear to the terrestrial-bound eye with such a contrast illustrates that there is a dilution of the contrast by other sources (zodiacal light and nightglow) in the night-sky light.

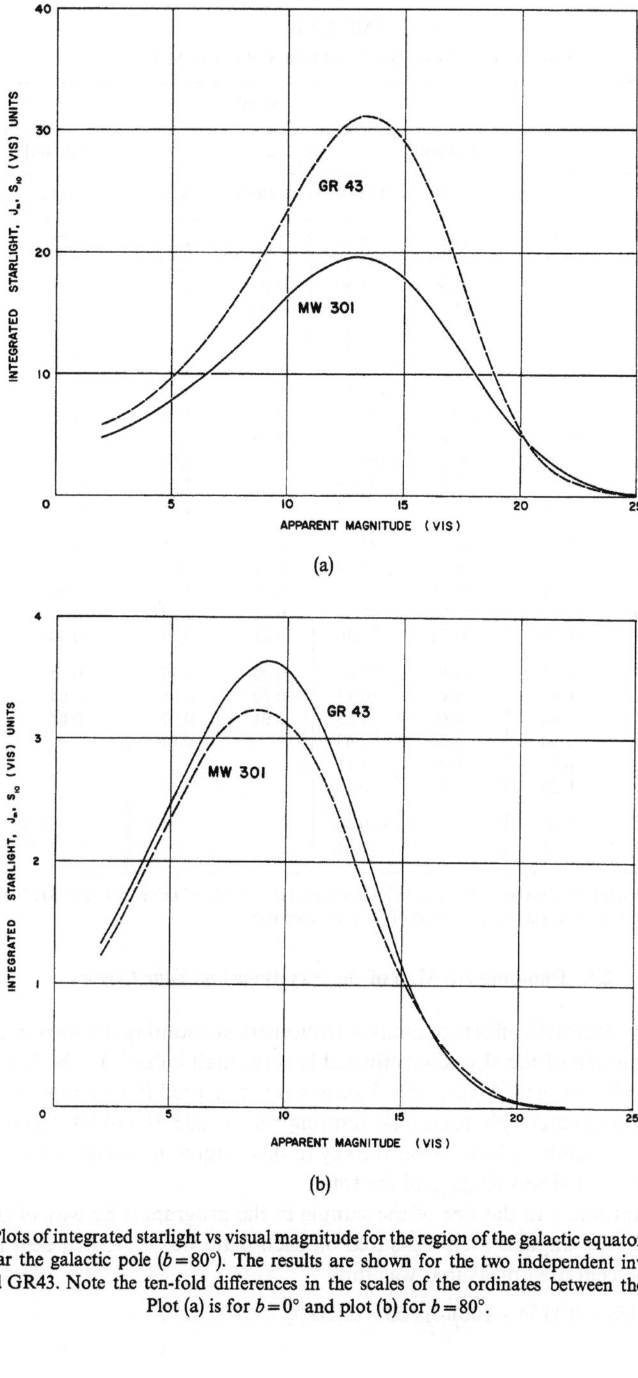

Fig. 2-7. Plots of integrated starlight vs visual magnitude for the region of the galactic equator ($b = 0°$) and a region near the galactic pole ($b = 80°$). The results are shown for the two independent investigations, MW301 and GR43. Note the ten-fold differences in the scales of the ordinates between the two plots. Plot (a) is for $b = 0°$ and plot (b) for $b = 80°$.

TABLE 2-II

Comparison of star-count data from GR43 and MW301[a].

m(vis)	b=0°				b=80°			
	J_m		J_m(cum)		J_m		J_m(cum)	
	GR43	MW301	GR43	MW301	GR43	MW301	GR43	MW301
2	5.91	4.61	341.63	238.12	1.33	1.23	37.29	34.04
3	6.73	5.42	335.72	233.51	1.67	1.59	35.96	32.81
4	7.88	6.42	328.99	228.09	2.07	1.98	34.29	31.22
5	9.42	7.65	321.11	221.67	2.49	2.38	32.22	29.24
6	11.37	9.10	311.69	214.02	2.90	2.75	29.73	26.86
7	13.81	10.76	300.32	204.92	3.26	3.04	26.83	24.11
8	16.70	12.60	286.51	194.16	3.52	3.22	23.57	21.07
9	19.98	14.52	269.81	181.56	3.63	3.25	20.05	17.85
10	23.45	16.37	249.83	167.04	3.56	3.12	16.42	14.60
11	26.77	17.98	226.38	150.67	3.29	2.84	12.86	11.48
12	29.47	19.12	199.61	132.69	2.85	2.44	9.57	8.64
13	31.03	19.58	170.14	113.57	2.30	1.98	6.72	6.20
14	30.96	19.21	139.11	93.99	1.73	1.51	4.42	4.22
15	29.00	17.93	108.15	74.78	1.19	1.08	2.69	2.71
16	25.24	15.84	78.15	56.85	0.74	0.72	1.50	1.63
17	20.20	13.15	53.91	41.01	0.42	0.44	0.76	0.91
18	14.68	10.19	33.71	27.86	0.21	0.25	0.34	0.47
19	9.55	7.32	19.03	17.67	0.09	0.13	0.13	0.22
20	5.46	4.82	9.48	10.35	0.03	0.06	0.04	0.09
21	2.66	2.88	4.02	5.53	0.01	0.02	0.01	0.03
22	1.05	1.54	1.36	2.65		0.01		0.01
23	0.29	0.72	0.31	1.11				
24	0.02	0.29	0.02	0.39				
25		0.09		0.10				
26		0.01		0.01				

[a] Entries corresponding to m(vis) > 20 for MW301 (Seares *et al.*, 1925) and m(vis) > 17 for GR43 (van Rhijn, 1925) are based on a quadratic extrapolation of the published data.

2.5. Photometric Map of the Sky Based on Star Counts

In spite of the dedicated efforts of many astronomers in counting the stars in Selected Areas, the fraction of the sky so scrutinized is very small indeed. In the Mt. Wilson program of 139 Selected Areas, 65683 stars were measured (an average of 472 per Area) to approximately photographic limiting magnitude 18 (16.94 visual magnitude). The total number of stars over the sky to this magnitude is about 3×10^8; thus the sample measured was 0.022% of the total.

Another approach to the size of the sample in the program is by way of the total area covered. The angular area measured on each plate was 0.1154 sq deg. For 139 Areas, the percent area measured was thus

$$(139 \times 0.1154 \times 100)/41\,253 = 0.039\%.$$

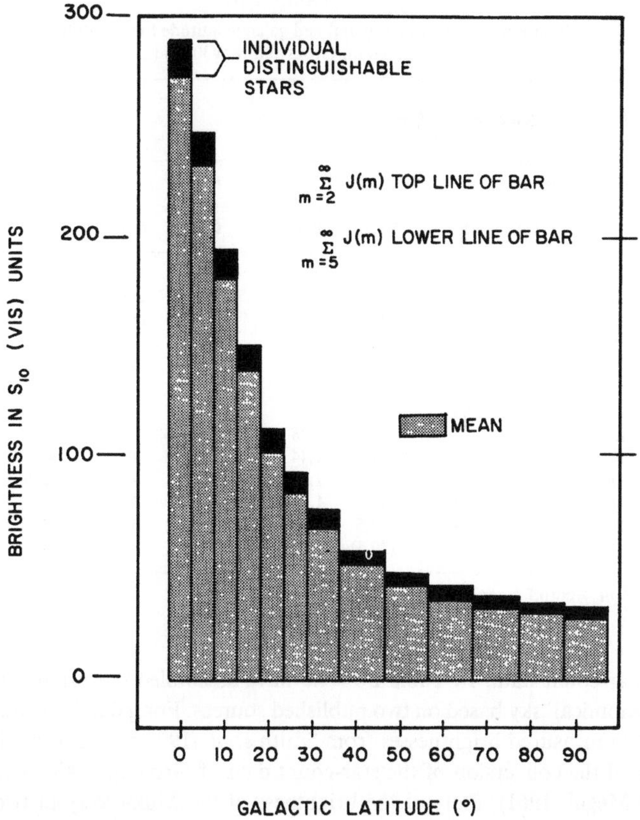

Fig. 2-8. Integrated starlight vs galactic latitude (mean of MW301 and GR43). The tops of the rectangles represent the total integrated starlight to magnitude 2 – the solid black portion; the shaded region corresponds to the total integrated starlight from stars fainter than those discernible to the eye.

From such a small sample we can hope to get only very crude morphological details of our galactic system and equally crude information on the photometric structure of the Milky Way. The uncertainty due to the smallness of the sample is compounded by the significant discrepancy between the two starcount reports quoted in this chapter.

The GR-43 listing is in the form of tables of the quantity $\log N_m$ from m(pg) 6 to 18 for every 10° of galactic longitude l and for galactic latitudes b of 0°, ±2°, ±5°, ±10°, ±15°, ±20°, ±30°, ±40°, ±50°, ±60°, ±70°, and ±80° (10296 entries for 792 regions in the sky). Since the study was based on 206 Selected Areas, it is obvious that considerable interpolation entered into the tabulated results.

In order to delineate the details in the regions of low galactic latitude, it is desirable to use photometric measurements rather than deduced brightnesses based on

TABLE 2-III

Relationship between J_m(cum) and galactic latitude for two different m_0,[a] based on GR43 and MW301 data

b	$J_m(\mathrm{cum}) = \sum_{m=m_0}^{\infty} J_m = J$							
	$m_0 = 5$				$m_0 = 2$			
	GR43	MW301	Mean	Relative	GR43	MW301	Mean	Relative
0	321	222	271.5	9.20	342	238	290	8.53
5	263	202	232.5	7.88	279	217	248	7.29
10	195	167	181	6.14	209	181	195	5.73
15	150	130	140	4.75	161	142	151.5	4.46
20	107	100	103.5	3.51	117	110	113.5	3.34
25		81	(82.5)	2.90		90	(92)	2.71
30	71	68	69.5	2.36	78	76	77	2.26
40	54	51	52.5	1.78	60	58	59	1.74
50	43	42	42.5	1.44	49	48	48.5	1.43
60	38	36	37	1.25	43	42	42.5	1.25
70		32	(33)	1.12		37	(38)	1.12
80	32	29	30.5	1.03	37	34	35.5	1.04
90		28	(29.5)	1.00		33	(34)	1.00

[a] m_0 = cutoff magnitude.

smoothed star-count data. In Table 2-IV we have assembled the brightness of the entire 'astronomical' sky based on two published sources. For galactic latitude $<30°$ we have taken measured brightnesses from Smith *et al.* (1970); for galactic latitudes $\geqslant 30°$ we used the conversion of the star-count data of Groningen 43 according to Roach and Megill (1961). Plots of the brightness of the Milky Way in two planes (Figures 2-9 and 2-10) illustrate its photometric structure, which is a result of our position in the Galaxy (compare with Figure 2-6).

2.6. The Total Light from the Stars

If the stars were uniformly distributed over the sky and diffused so that they were not individually discerned, they would produce a general glow that, according to Table 2-I and Figure 2-8 would be equivalent to 105 10th-mag. (visual) stars for each square degree of sky. In the hemisphere above the horizon there are 20626 sq deg, so that the total illumination from the sky is equivalent to 2.17×10^6 stars of visual magnitude 10.

Since a 10th-mag. star is about 100 times fainter than the faintest star visible to the average observer, alternative ways of expressing this are of interest. If, over the hemisphere, there were 51 stars as bright as the nearby and intrinsically bright star Sirius ($m = -1.58$), the total illumination would be the same as is now produced by the 30×10^9 stars in our Galaxy (Table 2-V). This is a dramatic indication of the facts that

Integrated starlight

l \ b[a]	Brightness J in S_{10}(vis) units																							
	0	2	5	10	15	19	30	40	50	60	70	80	−80	−70	−60	−50	−40	−30	−20	−15	−10	−5	−2	0
0	380	280	330	260	130	100	65	51	43	38	34	31	33	34	39	49	64	91	240	230	500	800	540	380
10	470	330	340	240	170	100	67	52	42	37	33	31	34	34	39	49	64	94	270	380	620	560	540	470
20	330	210	150	160	140	100	72	54	42	37	33	31	34	35	39	47	61	91	220	340	420	350	330	330
30	270	150	160	240	170	100	79	56	42	36	32	30	34	36	39	46	59	84	200	150	160	330	410	270
40	200	210	200	190	130	100	85	58	43	36	32	29	35	36	39	46	57	76	170	220	240	310	290	200
50	210	210	240	170	120	100	89	60	43	36	31	29	34	37	39	45	54	68	140	160	280	370	290	210
60	290	310	300	220	130	100	88	59	43	35	31	28	35	37	39	43	51	64	140	300	290	380	310	290
70	350	430	370	230	130	90	84	57	43	35	30	28	34	37	38	41	48	62	120	140	270	290	290	350
80	250	260	270	210	150	90	76	55	42	35	30	28	34	36	36	39	49	65	110	180	280	300	280	250
90	340	320	250	190	160	90	68	53	41	34	30	28	33	35	35	37	49	70	140	130	240	300	260	340
100	280	220	180	180	120	90	61	50	41	33	29	28	33	33	34	37	50	76	140	180	210	330	330	280
110	270	240	190	170	120	90	57	47	40	34	30	28	32	32	33	37	51	79	140	150	190	250	240	270
120	280	220	170	140	120	90	54	45	40	34	31	28	32	31	32	38	51	75	130	130	180	290	320	280
130	250	200	160	140	110	100	53	45	39	34	31	29	32	31	32	38	50	68	140	170	180	260	290	250
140	180	170	160	120	110	100	54	45	39	34	31	29	32	31	33	38	48	58	190	160	180	210	180	180
150	160	150	130	120	120	110	54	44	38	34	31	30	32	31	34	38	44	50	160	210	190	180	190	160
160	180	170	140	150	120	110	55	43	38	34	31	30	32	32	34	37	41	46	190	180	180	160	180	180
170	230	220	170	210	130	120	57	44	36	33	31	29	32	32	34	37	40	43	180	240	150	190	200	230
180	220	230	220	180	150	140	58	43	35	32	29	28	32	32	34	36	41	45	170	180	160	210	220	220
190	220	250	230	210	170	160	59	43	33	30	29	28	32	32	33	36	43	51	140	170	150	220	230	220
200	220	220	280	170	160	140	61	43	33	30	27	27	32	32	34	38	46	59	100	160	140	260	220	220
210	220	220	200	180	150	110	63	42	33	29	27	27	32	31	34	40	51	69	100	140	140	170	200	220
220	240	250	220	160	130	100	64	43	33	29	27	27	32	32	35	42	54	77	100	130	200	200	220	240
230	250	250	220	150	120	100	64	43	34	29	29	28	33	33	35	43	57	81	110	140	230	240	250	250
240	330	310	240	150	100	90	62	44	35	30	30	29	33	33	36	43	57	81	110	140	210	370	290	330
250	270	280	230	160	100	90	63	46	37	31	31	30	34	34	37	42	55	78	110	140	200	260	270	270
260	230	220	200	150	110	90	63	49	39	33	32	30	34	35	37	41	53	72	110	150	220	280	270	230
270	230	210	210	170	120	90	65	51	41	34	32	30	34	36	37	41	50	68	120	190	260	260	230	230
280	370	350	260	190	120	100	71	53	43	35	32	30	35	36	38	41	49	65	130	200	280	440	400	370
290	640	560	340	210	140	100	76	56	45	36	32	29	35	36	39	43	50	63	120	170	270	360	480	640
300	660	570	530	220	160	110	81	59	46	37	32	29	35	36	40	45	52	64	70	150	220	380	560	660
310	570	530	370	210	190	130	84	58	46	37	31	29	34	36	42	48	56	65	70	160	250	330	460	570
320	260	280	270	230	210	140	82	58	46	37	32	29	34	36	42	51	59	68	140	250	320	420	360	260
330	390	280	270	260	190	120	76	56	45	37	32	30	33	35	42	53	62	73	150	210	310	460	610	390
340	300	280	140	170	160	110	70	54	44	38	33	30	34	35	42	52	64	80	180	240	340	330	300	300
350	360	350	360	290	140	110	66	51	43	38	34	30	33	35	41	51	65	87	240	310	400	560	700	360
0	380	280	330	260	130	100	65	51	43	38	34	31	33	34	39	49	64	91	240	230	500	800	540	380

[a] $|b| \geqslant 30°$ from Roach and Megill (1961). $|b| < 30°$ from Smith *et al.* (1970). J at galactic pole = 28.

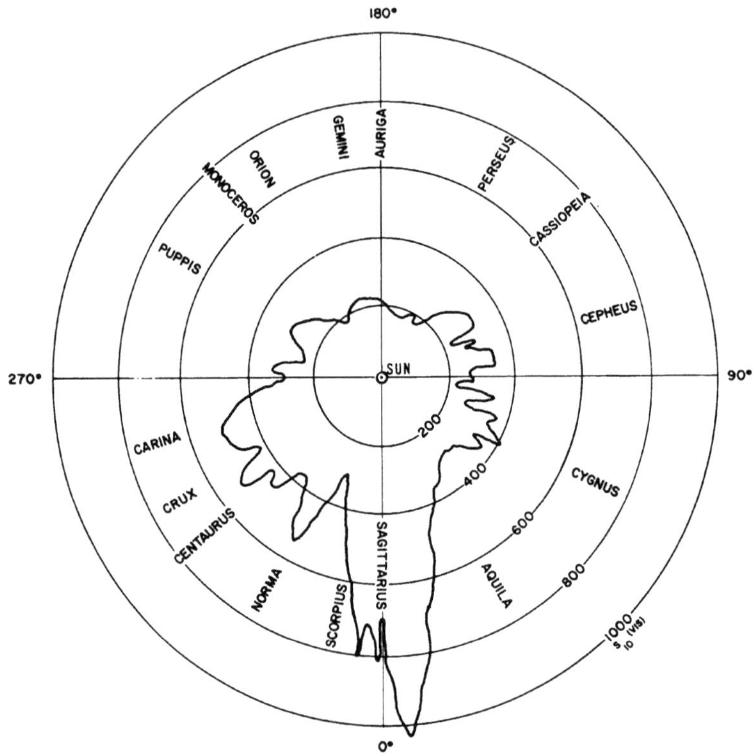

Fig. 2-9. Radial plot of Milky Way brightness for galactic latitude $-3°$. Galactic longitude is indicated increasing from 0° (toward the galactic center) in a counterclockwise sense.

most of the stars in our Galaxy are (1) much more distant than Sirius (2.7 pc $=8.8$ l-yr) and (2) intrinsically fainter.

The average surface brightness of the starlit sky can be expressed in terms of the surface brightness of the Sun, with its surface area of 0.2233 sq deg. In S_{10}(vis) units, the Sun's mean surface brightness is 2.18×10^{15}. Thus the surface brightness of the starlit sky is $105/2.18 \times 10^{15} = 0.5 \times 10^{-13}$ times that of the Sun.

As the realization grew that the stars are suns, a speculation arose in thoughtful minds concerning the possibility that the sky might, under some assumptions, approach a much higher brightness. Following an earlier suggestion by the astronomer Halley, Wilhelm Olbers, an Austrian physician, made the following statement in 1826:

> If there really be suns in the whole of space, and to infinity, and if they are placed at equal distances from each other, or grouped into systems like that of the Milky Way, their number must be infinite, and the whole vault of heaven should appear as bright as the Sun; for every line which may be supposed to emanate from our eye towards the sky, would necessarily meet a fixed star, and thus every point of the sky would bring to us a ray of sideral [*sic.*], or which is the same thing, of solar light. (Olbers, 1826.)

The discrepancy (or paradox, as it is sometimes called) between the speculative idea

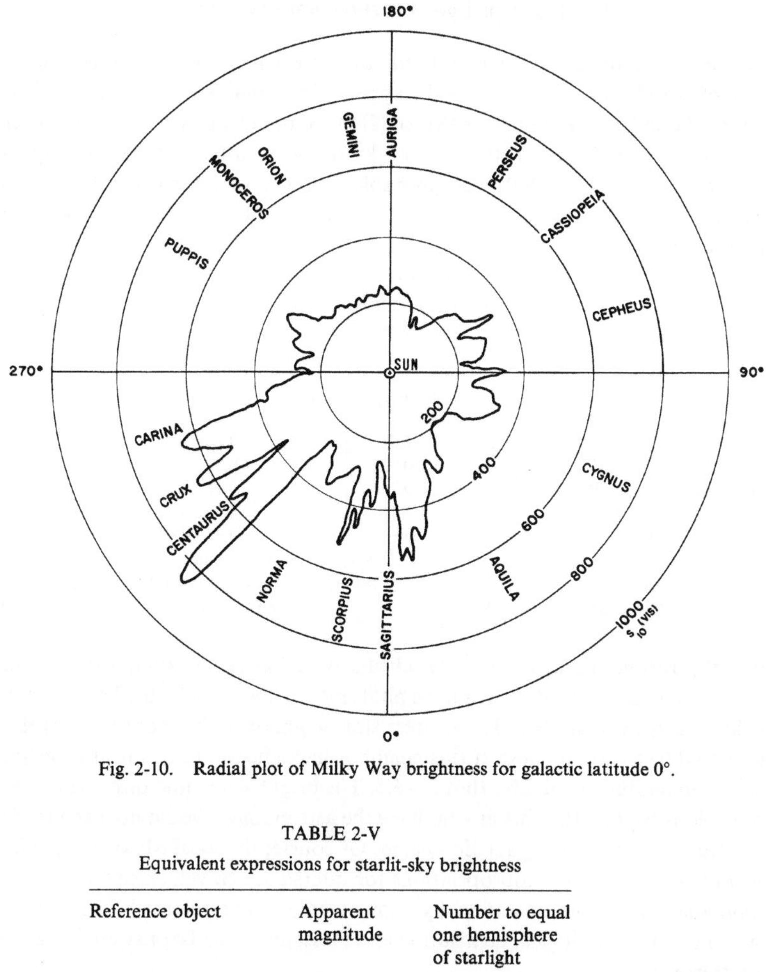

Fig. 2-10. Radial plot of Milky Way brightness for galactic latitude 0°.

TABLE 2-V

Equivalent expressions for starlit-sky brightness

Reference object	Apparent magnitude	Number to equal one hemisphere of starlight
10th-mag. star	10.0	217×10^4
1st-mag. star	1.0	590
Sirius	− 1.58	51
Venus (at brightest)	− 4.3	4
Full Moon	−12.5	22×10^{-5}
Sun	−26.72	445×10^{-12}

of Olbers and the observed 105 S_{10}(vis) is about 13 powers of 10. The Olbers' premise of an infinite number of stars extending in a uniform distribution to infinity is obviously not supported by the observations. The idea of continuing the discussion with galaxies replacing stars in the statement, however, has excited the interest of some cosmologists, and we shall return to this idea in the last chapter.

2.7. Our Felicitous Location in the Galaxy

We have called attention to our peripheral and somewhat isolated position in the Galaxy. How would the astronomical sky appear if we inhabited a planet associated with a star in (1) the center of our Galaxy, or (2) the center of a dense globular cluster?

In or near the galactic center, the general sky would be some 60 times as bright as our terrestrial sky (Table 2-VI). The probable presence of interstellar dust would have the effect of scattering the general starlight and would interdict any optical inquiries to the outside.

TABLE 2-VI

The night skies of different locations

Location	S[a]	$\bar{S}$		S_1		S_{10}(vis)	m(vis)
	(stars pc^{-3})	(pc)	(l-yr)	(pc)	(l-yr)		
Earth-Sun	0.1	2.15	7.03	1.19	3.89	105	–
Galactic center	45	0.28	0.92	0.16	0.51	6000	−4.1
Center of dense globular cluster	1000	0.1	0.33	0.055	0.18	48000	−6.5

[a] S is star density
$\bar{S}$ is mean separation of stars $=\sqrt[3]{S^{-1}}$
S_1 is the closest star neighbor $=0.554\,\bar{S}$
S_{10}(vis) is astronomical sky brightness
m(vis) is apparent visual magnitude of nearest star (assumed same luminosity as Sun)

A central position in a dense globular cluster would afford a brilliant sky indeed, but one in which the Milky Way would be photometrically overwhelmed by the many bright close stars over the sky. The nearest star neighbor, only about 66 light days distant, would be a brilliant star. If this neighbor happened to be a star like the Sun, its apparent magnitude would be about −6.5. The bright sky would discourage optical explorations beyond the cluster – perhaps the astronomers would not even realize that they were members of a galactic system. Of course, the local SPACE organizations would be busy raising appropriations for interstellar travel, since there is high likelihood that intercommunication by radio would occur relatively early in the development of a technological civilization and messages could be answered in about 132 days (terrestrial).

We consider that our particular location in the Galaxy is a happy one, even though we are certainly not suffering from stellar overcrowding. The advantage of a moderately luminous sky, which encourages rather than inhibits astronomical exploration, outweighs the loneliness of our position of relative isolation in the outer portion of our Galaxy.

Appendix 2-A. Star Gaging (Herschel) and Star Counts

A comparison of some of the numerical data of William Herschel's visual star gaging with modern star-counting results based on photographs will illustrate the nature of the statistics of stellar distributions over the sky.

Based on the discussion of Herschel's work in Hoskin (1963), we give a summary sample of some of his gaging results with the use of a telescope of 18.7-in. aperture and field of view of 15′ diameter (0.049 sq deg area).

Galactic latitude (deg)	Gage	N (sq deg)$^{-1}$
~ 0	79.0	1610
26	18.6	379
29	13.6	277
32	12.1	247
36	10.6	216
41	10.6	216
43	9.4	192
60	3.6	73
64	3.1	63
72	3.8	77
73	4.0	82

The numbers in the last column are the numbers of stars referred to a field of 1 sq deg and are therefore eligible for comparison with the quantity N_m (number of stars per sq deg brighter than magnitude m).

We show such a comparison in Figure 2-11, in which the solid lines refer to highly smoothed means from Groningen 43 (van Rhijn, 1925). We note that this sample of Herschel's observations suggests that his effective limiting magnitude was about 14 (visual). This value is reasonable considering the nature of his observing procedure,

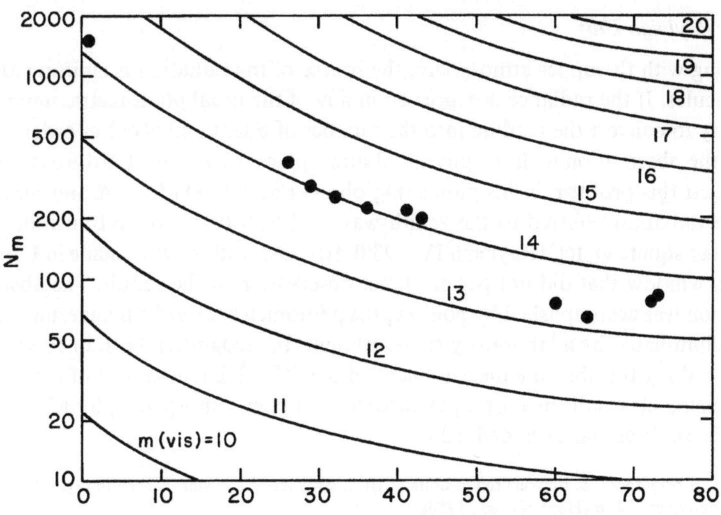

Fig. 2-11. Comparison of Herschel's star-gaging results with smoothed means from Groningen 43 modern star counts.

which necessarily involved a 'sliding field' instead of a stable field as in a modern automated telescope.

Appendix 2-B. Photometric Units in Surface Photometry

In this chapter we have used almost exclusively as a unit for expressing night-sky 'brightness', S_{10}, the equivalent number of stars of magnitude 10 per sq deg. In Chapter 4 we shall use another unit, the rayleigh, for indicating the quantal output of upper atmosphere emissions. To the general reader these units are probably unfamiliar and we shall discuss them here.

A. *The S_{10} Unit*

The use of the square degree as the unit of solid angle in the S_{10} unit follows historically from the fact that the square degree is the reference in stellar statistics (see Appendix 2-A, for example). There are 41 253 sq deg over a sphere, 3283 in a steradian – 1 sq deg is thus 0.000 305 ster. The choice of magnitude 10 rather than some other – such as 5, which is occasionally used – is arbitrary.

Except for the bright zodiacal light and the brightest part of the Milky Way, brightnesses in S_{10} units are expressible with three digits (less than 999), which is convenient in practical computations.

We have used either the astronomical visual or photographic magnitude scale in this book, avoiding the difficulties that arise in the use of the S_{10} unit in connection with the multicolor observations of modern photoelectric photometry.

B. *The Rayleigh Unit*

In dealing with the upper atmosphere, the source of the radiation is emitting atoms or molecules. If the radiance is expressed in any of the usual photometric units, it is necessary to convert the reading into the number of quanta involved in order to associate the observation with the physics of atmospheric processes. Lord Rayleigh IV recognized this problem in his pioneering observations of O I 5577 Å and reported that the radiation referred to the zenith was '... 1.81×10^{12} atomic transitions per second per square meter' (Rayleigh IV, 1930). His observations were made in London out of a window that did not permit direct observation of the zenith. The absolute calibration was accomplished by pointing the photometer toward a magnesium oxide screen illuminated by a laboratory standard light. In recognition both of Rayleigh's skill in making the absolute measurement of the 5577 Å intensity and of his insight in expressing his results in a form palatable to the upper-atmosphere physicist, a unit called the rayleigh has been defined –

If the surface brightness, B, is measured in units of 10^6 quanta cm^{-2} s^{-1} ster^{-1}, then in rayleighs, R, the surface brightness is $4\pi B$ (Hunten *et al.*, 1956).

Thus Rayleigh's measurement comes out as 181 R, in substantial agreement with much of the current data (Chapter 4).

The rayleigh unit is useful in upper-atmosphere research but is awkward in engineering applications because of the dependence of the quantal energy on wavelength. Also, in the case of emissions that are not monochromatic (for example, the nightglow 'continuum'), it becomes necessary to introduce a wavelength interval; this is often accomplished by the use of the rayleigh per ångstrom, R $Å^{-1}$.

The factor 4π in the definition can be interpreted intuitively as sweeping a photometer completely around a centimeter column of upper-atmosphere emission and capturing all the quanta from the column. Thus, if the column's effective thickness is 10 km (or 10^6 cm), as is approximately true in the lower ionosphere, then, recalling the 10^6 in the definition, the rayleigh brightness of a particular emission becomes the number of quanta (or atomic transitions) per cm^3 per s, which brings the investigator into direct quantitative contact with the basic physics of the nightglow phenomenon.

In Table 2-VII, we show the interrelationships of three alternative brightness units. Actually, the expressions brightness and intensity, both of which appear in the literature of the light of the night sky, are poorly chosen – better choices would be luminance and radiance. For a table giving the interrelationships among the luminance units commonly used by laboratory workers in photometry, see Walsh (1958).

TABLE 2-VII

Conversion factors for units of surface brightness at 5300 Å

Unit of surface brightness	S_{10}(vis)	R $Å^{-1}$	B (erg cm^{-2} s^{-1} $ster^{-1}$ $Å^{-1}$)
1 S_{10}(vis) (luminance)	1	4.40×10^{-3}	1.31×10^{-9}
1 R $Å^{-1}$ (radiance)	227	1	2.98×10^{-7}
1 B (quantal output)	7.62×10^8	3.35×10^6	1

References

Bok, B. J. and Bok, P. F.: 1968, *The Milky Way* (3rd ed.), Sky Publishing Corp., Cambridge, Mass.
Herschel, Sir William: 1802, *Phil. Trans.* 495.
Hoskin, M. A.: 1963, *William Herschel and the Construction of the Heavens*, Oldbourne History of Science Library, Oldbourne, London.
Hunten, D. M., Roach, F. E., and Chamberlain, J. W.: 1956, *J. Atmospheric Terrest. Phys.* **8**, 345.
Olbers, W.: 1826, *Edinburgh New Phil. J.* **1**, 141.
Rayleigh IV, Robert John Strutt: 1930, *Proc. Roy. Soc., London A* **129**, 458.
Roach, F. E. and Megill, L. R.: 1961, *Astrophys. J.* **133**, 228.
Seares, F. H., van Rhijn, P. J., Joyner, M. C., and Richmond, M. L.: 1925, *Astrophys. J.* **62**, 320 (Mount Wilson Publ. 301).
Smith, L. L., Roach, F. E., and Owen, R. W.: 1970, AEC Research and Development Report BNWL-1419, UC-2.
van Rhijn, P. J.: 1925, Publications of the Groningen Astronomical Observatory, No. 43.
Walsh, J. W. T.: 1958, *Photometry* (3rd ed.), Dover Publications, Inc., p. 520, App. III, Table II.

Sir William Herschel, 1785; by L. F. Abbott.
(Courtesy, National Portrait Gallery, London)

PLATE II

SIR WILLIAM HERSCHEL

1738–1822

Sir William Herschel, discoverer of the planet Uranus, was a versatile genius.

The first 20 years of his adult life were spent as a successful musician – organist, teacher, composer, conductor. During these years, however, he developed increasing interest in scientific matters – first through mathematics and its relationship to music theory, and then in astronomy and physics.

Although only an amateur astronomer in those years, he had carried out correspondence with recognized astronomers and written modest articles. Not until he discovered the new planet, Uranus, in 1781, was he accorded recognition as a professional, awarded the Copley Medal by the Royal Society, and elected a Fellow of the Society. Soon thereafter, he was appointed Astronomer Royal – the King's astronomer; this position gave him a permanent income and thus time and freedom to conduct research without having to depend on music for his livelihood. For the rest of his long life, William Herschel devoted his considerable talents to many facets of the field of astronomy.

Born in Hanover, Germany, on November 15, 1738, Frederick William was the second son of an oboist in the Royal Hanoverian Guards. Although uneducated by today's standards, the father encouraged his children in intellectual curiosity and love of nature. Having great interest in astronomy, he pointed out the features of the night sky to his children and discussed with them the work of such scientists as Kepler, Galileo, and Newton.

In 1755, William was sent to England with the Hanoverian Guards during the Seven Years' War. Although his health failed and he deserted to return to Germany, he was soon invited back to England to instruct a military band. His talents obtained him increasingly desirable situations until, in 1766, he became organist for the Octagon Chapel at Bath. In that locale, he became more and more interested in scientific matters – 'fluxions' (today's differential calculus), optics, physics, and astronomy. He developed considerable skill in devising optical instrumentation and refining current techniques, and devoted much of his night activities to observing the sky with instruments of his own making, one of which – the 40-ft-long reflector shown in the reproduction – was one of the largest telescopes of the time.

During this period, William began to be lonely for his family, so he invited his sister Carolina, 12 years younger than he and also unmarried, to join him in England, initially to help him as a singer in his musical activities. With his increasing devotion to studies of the night skies, however, she became adept at assisting him in that as well, recording his observations and so freeing him from the necessity of moving his eyes from their intent search for new and wonderful discoveries.

Forty-foot telescope designed, constructed, and used by Herschel at Slough.

Inasmuch as she had been working closely with William for several years before his discovery of the 'comet' that turned out to be Uranus, it is probable that she took the notes on the night that he made that significant observation. Her observational abilities developed sufficiently that, over the years, she discovered eight comets herself, with the 7-ft Newtonian sweeper that William gave her.

In 1788, when he was 50, William Herschel married a much younger English woman, by whom he had a son, John, who also became an eminent scientist. Although the marriage first estranged Carolina, William's wife was consistently friendly to her, and the birth of their son restored the close family bonds.

Although his discovery of Uranus was the feat that made him famous, William Herschel made other major contributions to the field of astronomy that are of equal, and possibly even greater scientific value – the refinement of telescopic optics; his analysis of the 'structure of the heavens', the forms of the Galaxy and nebulae, and a theory of the continuing process of their formation, from 'birth' through stages of development and growth to decline, old age, and 'death'; the period of rotation of Mars, to within 5 s of the value accepted today; his theory of self-luminous material, usually surrounding specific stars, as potential future stars; and numerous others, such as his early work on star counting.

Working with a 20-ft Newtonian reflector of 17.8-in. aperture, Herschel embarked on a long-range program of star-density estimation, which he termed 'star-gaging'. Directing his telescope to a particular part of the sky, he counted the stars visible in the field at each setting of the instrument, then moved to a nearby field and repeated the process. He initially assumed uniform distribution of the stars throughout space but, as he continued to compile data he abandoned this view, commenting in 1802, '... this immense starry aggregation is by no means uniform. The stars of which it is composed are very unequally scattered, and show evident marks of clustering together into many separate allotments' (*Phil. Trans.* 1802:495). This work, carried out over many years, became the basis of the star counts ably carried on and refined by later astronomers such as Kapteyn and Bok.

In this day of specialization in all fields, including astronomy, it is challenging to realize the tremendous scope, breadth, and sophistication of intellectual investigation this completely self-educated man was able to bring to bear on his chosen field of science.

William Herschel was truly a giant among astronomers – a first-magnitude star in his own Galaxy.

Professor J. C. Kapteyn, 1919:
photograph taken by F. Ellerman.
(Courtesy, Mount Wilson Observatory)

PLATE III

JACOBUS CORNELIUS KAPTEYN

1851–1922

It was not until more than a century after Herschel wrote '... this immense starry aggregation is by no means uniform' (*Phil. Trans.* 1802:495), that the Dutch astronomer Jacobus Cornelius Kapteyn devised an appreciably more refined technique for determining the numbers and distribution of stars in the heavens.

As one of fifteen children of the director of a boarding school in Barneveld, The Netherlands, Kapteyn's aptitude for science had been nurtured since early childhood. Not only Jacobus but several of his brothers became leaders in science. He obtained an excellent education at the University of Utrecht, where he received his doctoral degree in physics in 1875. The only position available at the time, however, was that of observer at the Leiden Observatory, and he accepted that. As a consequence, his scientific talents were bent toward astronomy – to physics' loss! Three years later, at age 27 – very young for the honor – he was named full professor in astronomy at the University of Groningen. Shortly thereafter he married Catharine Elizabeth Kalshovan, by whom he had two daughters and a son.

Despite the honor of the professorship, Kapteyn was frustrated because the University of Groningen had no observatory or observing instruments. Like Herschel, he was interested in the structure of the Universe (Galaxy) and to study that, he needed observing instruments. Fortunately, he secured permission to use the meridian circle of the Leiden Observatory during a professor's vacations. On this instrument, he planned a careful program for determining stellar parallaxes, introducing the differential method of observations in right ascension. To that time, the parallaxes of only 34 stars had been obtained, the most effective means being an instrument called a heliometer, and it was a slow technique.

Kapteyn's new approach yielded parallax information of accuracy comparable with that of the heliometer, and more rapidly, but was still not speedy enough for his requirements. By 1889 he had devised an ingenious photographic technique for measuring the parallaxes of large numbers of stars quickly and accurately.

The general problem of establishing the structure of the Galaxy was furthered by his speeded-up methods, but still was obscure because of lack of data. He soon realized two important points – first, by measuring proper motions of stars, he could determine their parallaxes with respect to the ever-increasing base line of the Sun's motion through space, and second, for the structure of the universe, it was not necessary to know the distances of individual stars, but rather the mean distances of groups of stars at different galactic latitudes.

Throughout his life, he adhered to a highly practical approach to scientific investigation. He could visualize the overall aspects of a major problem and, at the same time, develop techniques for utilizing limited available information and work out ways of obtaining additional and better data and refining the analysis and application of the new data to the problem. He focused his life's efforts on the galactic-structure problem, and this ability to expand and refine at the same time, iteratively, was an invaluable asset.

Kapteyn's investigations on proper motion led to his discovery of two star streams. Instead of random motion of stars, as had long been assumed, he found that the distribution of motion consistently tended in two preferential directions. Further, the lines seemed to converge on two points – one just south of α Orionis, the other just south of η Sagittarii. These findings convinced him that these two star streams were parallel to the lines joining our own solar system with the two apparent end points.

Such a concept necessarily influenced the then-current ideas about the structure of the Galaxy. But immense quantities of additional data, from far-flung sources, were needed to provide more reliable measurements and data for fainter stars.

At this point Kapteyn's genius for organizing the meticulous details necessary for a concerted attack on this major problem came to the rescue. He initiated steps to establish an international network of cooperative observing institutions. To systematize the monumental task of obtaining the desired data, in 1906 he published what has become known as his famous Plan of Selected Areas – 206 carefully selected regions over the entire celestial sphere – in which the observers would secure measurements on magnitude, proper motion, parallax, class of spectrum, and radial velocity for stars recorded on graded photographs centered on the selected regions.

In those days of limited, slow travel and communication, such a project might well have seemed impossible to many scientists. Kapteyn, however, being a gregarious man who had traveled widely, enjoyed the friendship as well as the professional respect of astronomers all around the world. This wide acquaintance made possible the tremendous coordinated activity of collecting data on his Selected Areas. By the time of his death, in 1922, the data that had poured in from all over the world had made possible the highly refined picture of the structure of our Galaxy that was essentially what we have used until very recently – more than 50 years after he conceived the project.

As in most areas of science, the solution of one problem brings up a host of new questions. And this has been true of Kapteyn's galactic structure – in the past decade radio astronomers have begun to find anomalies in the picture and have initiated new directions of investigation with their own new arsenal of research tools.

Kapteyn's contributions to our knowledge of the structure of the universe brought him well-deserved honors from all over the world.

Bart J. Bok with Telescope at Kitt Peak, Arizona.
(Courtesy, B. J. Bok)

PLATE IV

BART JAN BOK

1906–

Holland has contributed more than its share of outstanding astronomers in the past century, particularly in the area of galactic-structure research. In addition to Kapteyn, another Netherlander, Bart Jan Bok, has further refined the counting approach and contributed to a more precise shape-and-size description of our Galaxy than was possible before his time.

Born in Hoorn, The Netherlands, in April 1906, Bok studied at Leiden University. Although he studied there several years after Kapteyn's death, the influence of the Selected Areas proponent was still strong and undoubtedly served as inspiration to many of the young students, including Bok, who completed his studies there in 1927.

At the 1928 General Assembly of the International Astronomical Union, held in Leiden, Bok was on the Reception Committee, where he chanced to meet a young astronomer from Smith College, Priscilla Fairfield. After a year's courtship, he persuaded her to marry him, and they moved to Harvard University.

Harvard had a strong Astronomy Department, filled with eager and innovative graduate students and young professors, and Bok fitted in with ease, enjoying the stimulating atmosphere and associates. He obtained his American citizenship in 1938.

During his long career at Harvard – 1929 to 1957 – Bok held the G. R. Agassiz and R. W. Willson fellowships in the early years, became assistant professor in 1933, associate professor in 1939, and Robert Wheeler Willson professor from 1946 to 1957. He served as Associate Director of Harvard Observatory from 1946 to 1952.

Even in those busy years, he found time for research in his own areas of interest – galactic structure, in particular. One of his early researches, in the 1930s, was conducted on photographs of the constellation Carina in the southern sky – visible beyond the 'upper edge' of the Sagittarius arm of our Galaxy (see Figure 2-6). Some of the results of that study contributed to the development of his picture of the Galaxy as given in his book, *The Distribution of the Stars in Space*, published in 1937, which covered the state of knowledge at that time on the basis of the earlier work of Kapteyn, van Rhijn, and Seares, as well as of Bok and his students.

Astronomy students at Harvard in those early years counted millions of stars under Bok's tutelage, and he then placed many of them in strategically located departments of astronomy in the southern- and mid-U.S. universities. Each member of this 'star-counting brigade' was responsible for a different section of the Milky Way. The original purpose of the brigade was to define the structure of our Galaxy in the general vicinity of our solar system, but it soon outgrew that limitation. With the tremendous quantities of data generated, it became possible to fill in many of the gaps in the then very sketchy picture of our Galaxy. Gradually the emphasis shifted from studies of local stellar distribution to investigations relating to the spiral structure of our Galaxy. Additional inputs in the past decade, from such newcomers to the field as the radio astronomers, permit us to assume with relative certainty a rotating spiral structure with several arms, at least three of which have sections in Earth's quadrant of the Galaxy (Figure 2-6).

In the 1940s, Bok initiated concentrated investigation of globules, the smallest dark nebulae, obtaining results that led him to stress their importance as protostars. Later work, by both him and other researchers, has confirmed and expanded this concept, which has become important in studies of the life cycles of stars and star clusters.

In 1957, Bok accepted the position of Head of the Department of Astronomy at the Australian National University in Canberra and was also named Director of the Mount Stromlo Observatory. In Australia, he concentrated especially on research of the brilliant and beautiful Magellanic Clouds, appendages to the galactic structure, and on work relating to the spiral structure of the Carina Section of the Southern Milky Way.

He continued this line of research after returning to the United States, where he was appointed Head of the Department of Astronomy at the University of Arizona and Director of Steward Observatory in 1966. From that base, he was able to do considerable work on the spiral structure of the southern Milky Way using the telescopes of Cerro-Tololo Inter-American Observatory in Chile. Working in close collaboration with radio astronomers and theoretical workers in the field, Bok and his associates (principal among them, Priscilla F. Bok) have extended our knowledge of southern spiral structure in the Carina-Centaurus Section to distances up to 8000 pc from the Sun. Kinematically and structurally, the Carina Feature is now one of the best-known spiral features of our Galaxy.

Two key elements have characterized Bok's long career in astronomy. First, he has maintained meticulous use of statistical methods in studying the galactic structure and interstellar matter; this tenacious devotion to obtaining and analyzing massive quantities of data on which to base a concept – and of continuing to obtain and analyze data to confirm, refine, and expand the concept – has put him in the enviable position of being able to stand solidly on early statements and use them as foundations for future work. Second, his wide range of interests and lively curiosity about new questions that arise in solutions of old problems have kept him in the vanguard of 20th-century astronomy.

Since retiring from his official posts at the University of Arizona in 1970, Bart Bok and his wife Priscilla have completed the revision of their *The Milky Way* (1968; first published in 1938). He continues to publish articles on past research, and there are still those new questions that lure him into fresh research efforts.

Bart Bok has made, and continues to make, major contributions to astronomy generally, and particularly to the field of galactic structure, on which he has focused the spotlight during the greater part of his long professional life. And he is still going strong!

CHAPTER 3

THE ZODIACAL LIGHT AND GEGENSCHEIN

The projection of the Earth's orbit on the celestial sphere, the ecliptic, delineates the apparent annual path of the Sun against the star background. The other major members of the solar system – the planets and their moons – move within a few degrees of the ecliptic as a consequence of their intrinsic orbital motions combined with that of the Earth.

Prehistory does not, of course, document the first of our progenitors who first took advantage of his new, erect position to look inquisitively at the sky, noting first the Moon's steady march among the stars as it went through its monthly waxing and waning, then that other bright objects (the planets) wandered more slowly against the same general stellar background.

The ancients of different civilizations seem to have independently animated the ecliptic circle with 'living beings', which led to the Greek word *zōidiakos* and our present zodiac for the region along and near the ecliptic. The ancient lore of the zodiac included not only the imaginative outlines of the creatures of the 'zoo' into constellations (Figure 3-1) but also a whole library of tales of the activities, godly and ungodly, of the characters of mythology who traversed the zoological garden above the anxious eyes of the ancients and in turn served as their sometimes genial and sometimes stern chaperones.

That a relatively minor component of the solar system, a diffuse cloud of dust grains, also favors the ecliptic (zodiacal) region is evidenced by the existence of a photometric brightening toward the ecliptic due to the scattering of sunlight by the dust. The optical phenomenon carries the name of zodiacal light, and the physical that of the zodiacal cloud.

In the absence of moonlight and the aurora, the zodiacal light is the brightest component of the night sky. That it is not generally recognized is probably due to a number of factors. For example:

(1) It is best observed near the western horizon after twilight and near the eastern horizon before dawn, and is often confused with the twilight/dawn phenomena.

(2) Because it is brightest near the horizon, the potential zodiacal-light observer suffers from the competition of low-lying clouds and city lights.

(3) Since the ecliptic in mid and high latitudes is often significantly inclined toward the horizon, the zodiacal light is best seen in the tropics, where the ecliptic is more or less vertical with respect to the horizon.

For a well-located observer, the zodiacal light is a splendid spectacle (Frontispiece), appearing visually as a cone of light oriented along the ecliptic and brightest at the

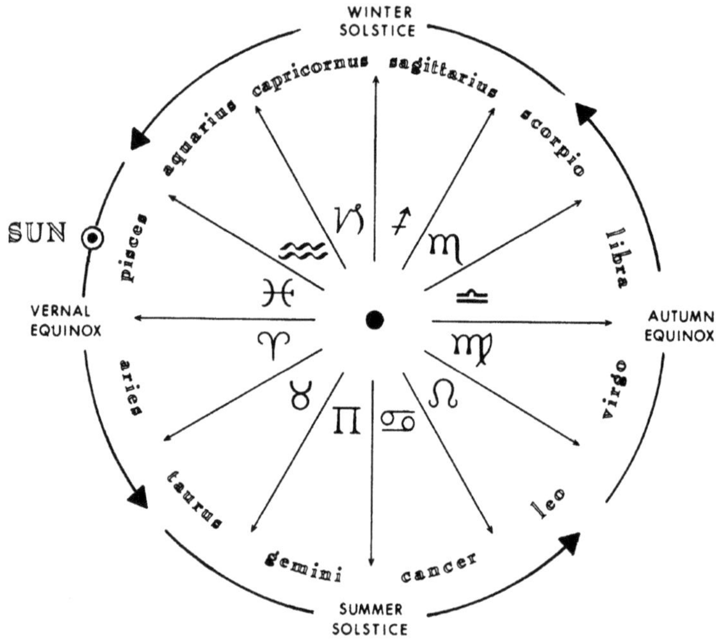

Fig. 3-1. Circular plot of the Sun's annual traverse among the 12 constellations of the zodiac, in the direction of the arrows. The zero point for the measurement of right ascension and of ecliptic longitude is at the intersection of the ecliptic and the celestial equator (the vernal equinox). In ancient times (Hipparchus, 2nd century B.C.), this intersection occurred in the constellation Aries but, as a result of the precession of the Earth's orbit, it now occurs in the preceding constellation, Pisces. It will move clockwise through one sign each approximately 2000 years, arriving in Aquarius, for example, in about 4000 A.D.

horizon. It diminishes in both brightness and width as it extends well up toward the zenith. Some observers have even reported visually detecting the zodiacal (ecliptic) brightening along the entire ecliptic circle. At its brightest, as seen from the Earth's surface, it is some three times as bright as the conspicuous southern Milky Way. At its faintest, it is two to three times as bright as the integrated starlight at the galactic pole.

Figure 3-2 shows photometric slices across the zodiacal light based on the tabulation in Appendix 3-A. At an elongation of 30° from the Sun, the width of the cone to the visual observer is about 60° but, according to photometric measurements, the zodiacal light extends out to the ecliptic pole, where its brightness – c. 78 S_{10}(vis) – is a little less than that of the average integrated starlight – c. 105 S_{10}(vis) – and comparable with the nightglow continuum (see Chapter 4). This near equality of the three components compounds the difficulty of their disentanglement. The apparent visual width of the cone decreases at greater elongations, and the slices shown in Figure 3-2a include most of the domain easily observable to the eye. Photometric measurements confirm that there is an ecliptic enhancement at all elongation angles (Figure 3-2b).

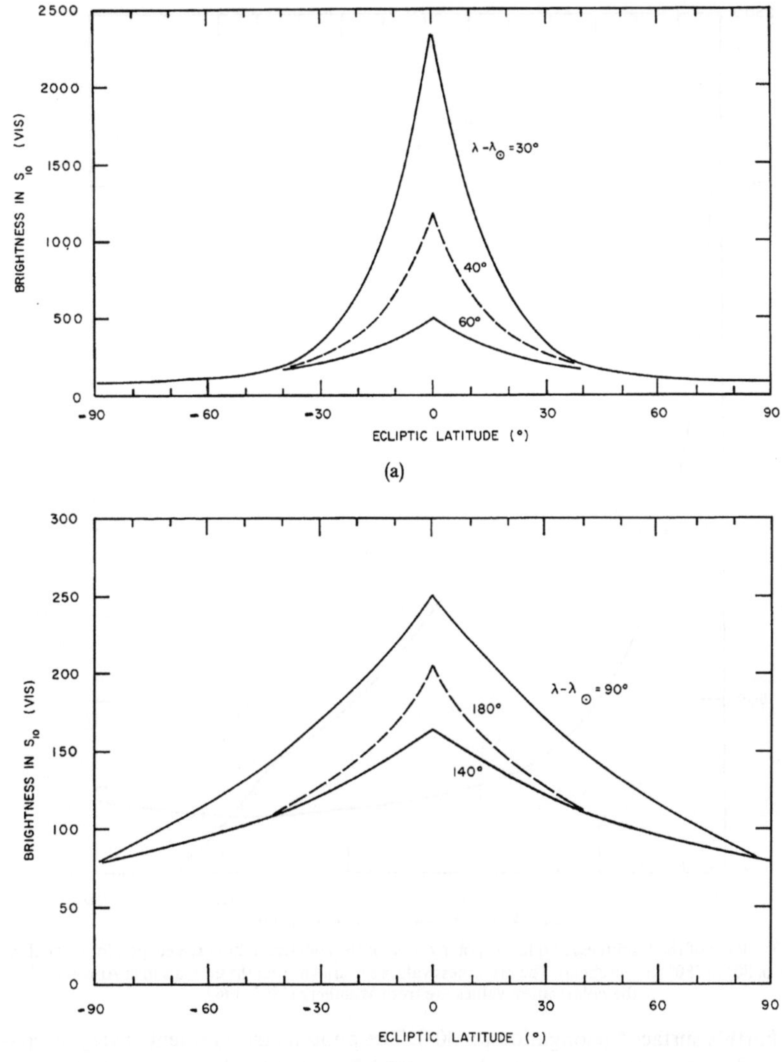

Fig. 3-2. Photometric slices across the zodiacal light, based on the model in Appendix 3-A. Note that there is a 10-fold change of ordinate scale between Figure 3-2a (the bright region relatively near the Sun) and Figure 3-2b (the fainter region at high solar elongation angles). The slight increase in the brightness in the ecliptic from 140° to 180° is due to the gegenschein (see text).

The variation of zodiacal-light brightness along the ecliptic has been extensively studied. Figure 3-3 shows the brightness for that part of the phenomenon observable

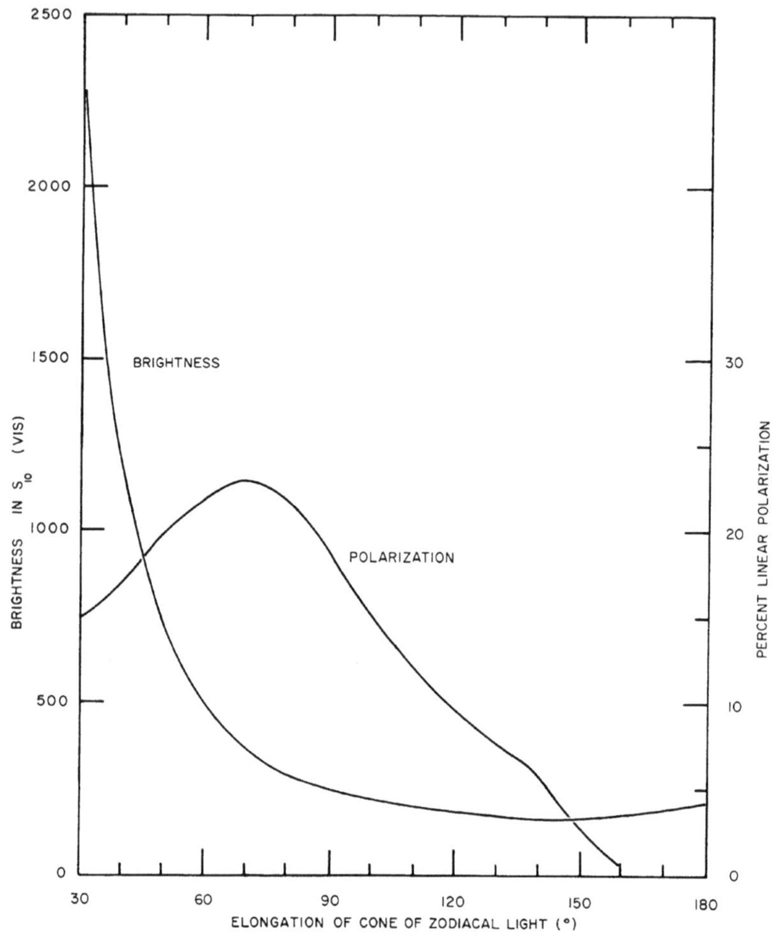

Fig. 3-3. Plots of the brightness and linear polarization of the zodiacal light in the ecliptic for solar elongations of 30° to 180° (gegenschein). The brightness values are drawn from the tabulation in Appendix 3-A; the polarization values are from Weinberg (1963, 1964).

from Earth's surface* (elongation $\varepsilon > 30°$). The photometric gradient is very steep – from $\varepsilon = 30°$ to $\varepsilon = 45°$ along the ecliptic, the differential brightness is 1385 S_{10}(vis), almost the same as the difference at $\varepsilon = 30°$ from the ecliptic ($\beta = 0°$) to $\beta = 15°$, 1449 S_{10}(vis). These sharp photometric gradients give the visual impression of a definite edge to the phenomenon. Note in Figure 3-3 that the steep, exponential-type rise for the smaller elongations plotted suggests that the brightness of the zodiacal

* The zodiacal light can be observed without twilight contamination after astronomical twilight or before astronomical dawn, when the Sun is more than 18° below the horizon. Most observers have limited their measurements to elevations at least 10° above the horizon. Thus, if the ecliptic is vertical for the observer, the closest observation to the Sun is 28°.

light close to the Sun (angle-wise) must be very conspicuous. This point was brought out by van de Hulst (1947), who called attention to the fact that the terrestrial observations for $\varepsilon > 30°$ could be extended to the observations of the F (Frauenhofer) component of the solar corona on a log-brightness/log-elongation plot by a straight line with a slope of -2.4. Thus,

$$\log B = \text{const.} - 2.4 \log \varepsilon.$$

That there are two components of the solar corona was established by Grotrian (1934). One, labeled K, is the true corona, with streamer-like structures so often photographed at times of total solar eclipse. The other, the F corona, presumably due to sunlight scattered by dust in the line of sight, close to the Sun (in angle, but not physically), is less than 0.1 as bright as the K corona near the Sun's limb but dominates photometrically for elongations greater than one solar radius from the solar limb. It is this latter component of the corona that is associated with the zodiacal light.

In Figure 3-4 we have plotted the log of the zodiacal-light ecliptic brightness (from

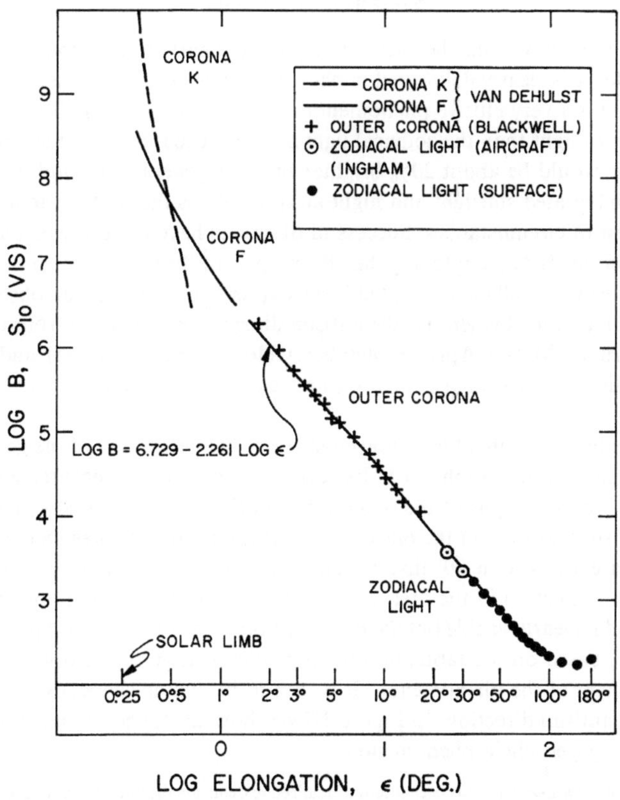

Fig. 3-4. Log ecliptic brightness of the zodiacal light against log solar elongation (Roach, 1972).

Appendix 3-A) against log elongation angle. The slope of the dashed line connecting the faint end of the F corona (1°.25 from the Sun's center) and the ground observation at $\lambda - \lambda_{\odot} = 30°$ is -2.37, in agreement with van de Hulst's early interpolation.

3.1. Polarization of the Zodiacal Light

The accurate quantitative measurement of the linear polarization of the zodiacal light is difficult because the direct observation is necessarily of the composite light of all the components of the light of the night sky, and in order to refer the measured polarization to the zodiacal light it is necessary to make arithmetic allowance for the dilution from the integrated starlight and airglow. In Figure 3-3 we include a plot of the linear polarization of the zodiacal light in the ecliptic plane according to measurements by Weinberg (1963). Over the elongation range 30° to 160° there is a well-defined maximum of 23% at 70°.

3.2. The Gegenschein

An extended faint glow* may be seen centered on that part of the ecliptic opposite the Sun. Its existence is clearly delineated in photometric measurements (see Figures 3-2b and 3-3) but it is a very elusive phenomenon to the visual observer. If there were no other sources of light than the zodiacal light and the gegenschein, the antisun region of the ecliptic would be about 20% brighter than the general surroundings. But the presence of integrated starlight and night airglow dilutes the contrast to about 10% under the best of circumstances. Successful observers have usually swept across the sky using averted vision, employing the off-axis part of the retina, which is especially sensitive if the eye is well dark-adapted (see Chapter 1, p. 3). Optimum observational conditions are obtained when: (1) the antisun direction is well away from the Milky Way (in February, March, April; September, October, and November), and (2) when the antisun direction is as high as possible above the horizon – that is, near local midnight.

To some extent the difficulties involved in the visual perception of the gegenschein are reflected in its photography or in its accurate photometric measurement. Figure 3-5 shows two photographs of the same region of the astronomical sky made over a three-month span so that, in the one case, the antisun direction was included and in the second case it was some 90° distant. The presence of a glow more or less centered on the antisun direction in the first case is apparent. In the second photograph, the zodiacal band is clearly visible but the bright region of the first photograph is missing.

Figure 3-6, based on the tabulation in Appendix 3-A, gives in isophote form the shape and extent of the phenomenon. It is seen to be oval in shape, extending some 40° from the antisun direction. In Figure 3-7 we show an isophotal map of the entire zodiacal-light/gegenschein phenomenon.

* The German word *gegenschein* has been internationalized to delineate the phenomenon: *gegen* (counter) *schein* (glow).

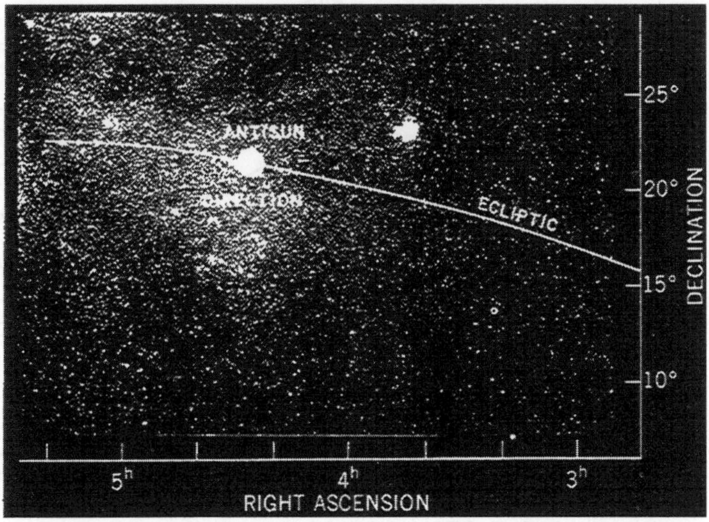

(a)

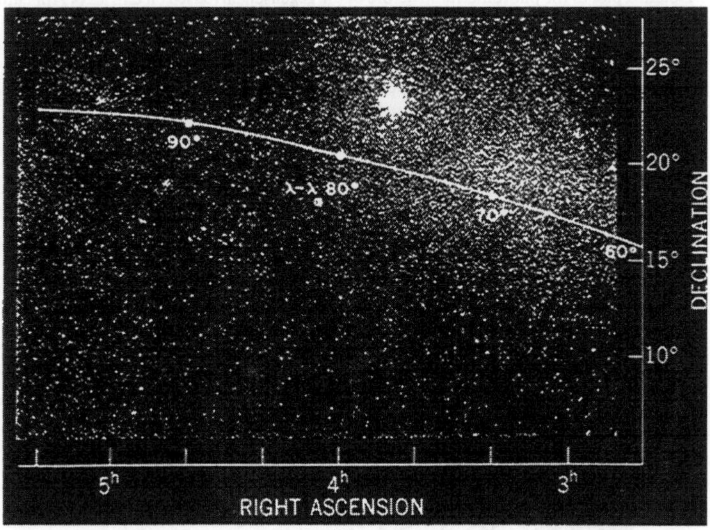

(b)

(Courtesy R. G. Roosen, McDonald Observatory)

Fig. 3-5. Two photographs of the same astronomical region of the sky, taken three months apart. (a) Taken on December 1, 1966, this photograph shows the gegenschein as a general patch of light centered near the antisun direction. (b) Made on March 3, 1967, when the antisun direction was some 90° to the east, this photograph shows a part of the zodiacal band from elongations 60° to 80°; the gegenschein is, of course, no longer in the field of view.

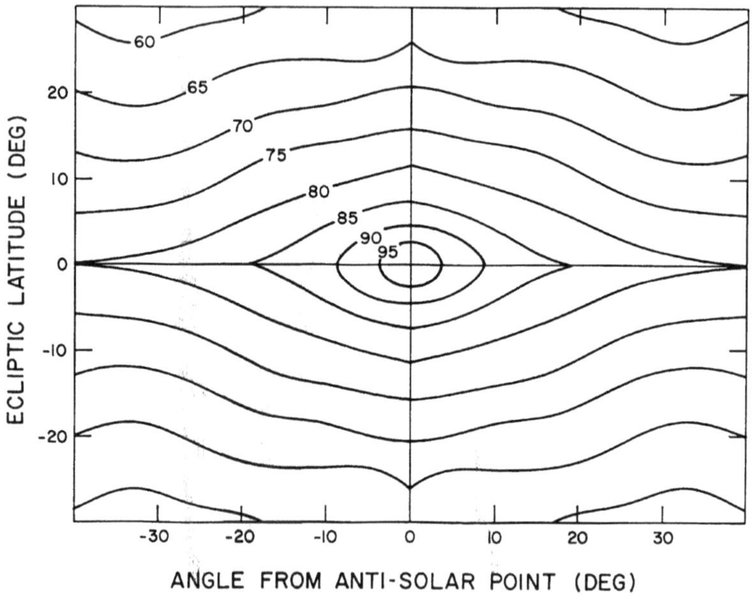

Fig. 3-6. Detailed plot of the photometric structure of the gegenschein (based on entries of Appendix 3-A), referred to relative brightness in percent of antisun value – 205 S_{10}(vis).

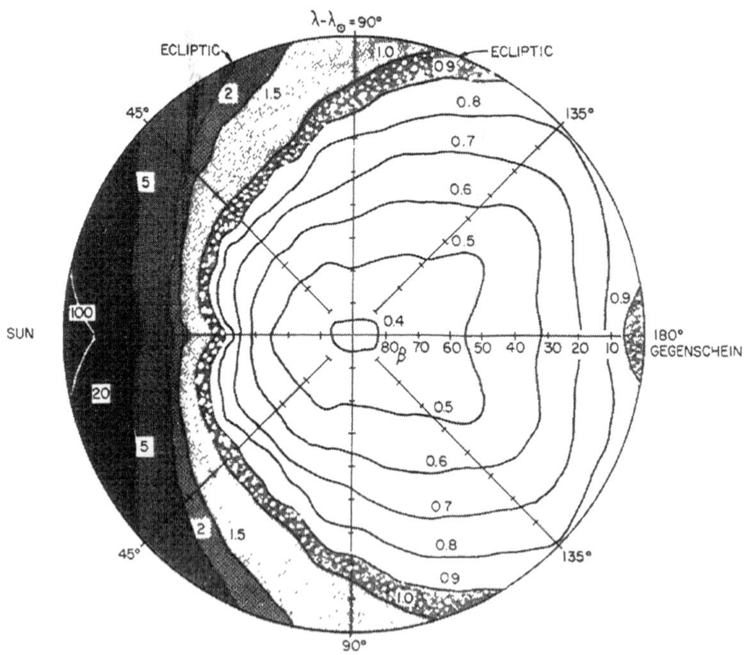

Fig. 3-7. Isophotal map of entire zodiacal light in polar-coordinate system (based on Appendix 3-A). Outer circle is ecliptic; center is ecliptic pole. Ecliptic latitudes (β) are indicated. The brightness unit is 1.0 for brightness in antisun (gegenschein) direction.

In Chapter 6, we shall return to the problem of the zodiacal light and the gegenschein, with particular reference to the scattering of sunlight by interplanetary dust. We shall there discuss not only the general interplanetary dust cloud but also its possible perturbations in time and space.

References

Grotrian, W.: 1934, *Z. Astrophys.* **8**, 124.
Roach, F. E.: 1972, *Astron. J.* **77**, 887.
van de Hulst, H. C.: 1947, *Astrophys. J.* **105**, 471.
Weinberg, J. L.: 1963, *Photoelectric Polarimetry of the Zodiacal Light at λ5300*. Ph. D. Thesis, Univ. of Colorado.
Weinberg, J. L.: 1964, *Ann. Astrophys.* **27**, 718.

APPENDIX 3-A

The zodiacal light and gegenschein

Numbers in ecliptic ($\beta=0$) taken from Figure 3-4. Off-ecliptic numbers (Roach, 1972)

$\lambda-\lambda_\odot$	Values of brightness in zodiacal light and gegenschein (S_{10}(vis) units)																			$\lambda-\lambda_g$
	$\beta=0$	5	10	15	20	25	30	35	40	45	50	55	60	65	70	75	80	85	90	
180	205	182	168	156	145	135	126	119	112	106	101	97	93	89	87	84	82	80	78	0
175	191	176	165	154	144	130	125	117	113	110	106	102	98	94	90	86	83	80	78	5
170	183	173	162	152	140	132	124	119	114	109	104	101	97	93	90	86	83	80	78	10
165	178	169	158	150	139	132	124	118	113	108	103	101	97	93	90	86	83	80	78	15
160	173	165	154	147	138	129	121	115	109	105	101	99	95	92	89	86	83	80	78	20
155	169	161	151	141	136	128	120	113	106	103	100	97	94	91	88	86	83	80	78	25
150	167	158	149	139	131	124	117	111	105	102	99	96	93	90	88	85	83	80	78	30
145	165	156	147	139	132	125	118	112	105	102	99	96	93	91	88	85	83	80	78	35
140	164	155	147	140	133	127	121	115	109	106	102	99	95	92	89	86	83	81	78	40
135	165	157	151	145	139	133	128	123	118	113	109	104	100	96	92	88	85	81	78	45
130	173	166	158	151	145	139	133	127	121	116	111	106	102	97	93	89	85	82	78	50
125	177	169	162	154	148	141	135	129	123	117	112	107	102	99	94	89	85	82	78	55
120	190	181	172	164	156	148	141	134	128	122	116	110	105	100	95	90	86	82	78	60
115	191	182	173	165	157	149	142	135	129	122	116	111	105	100	95	91	86	82	78	65
110	200	190	180	171	162	154	146	139	132	125	119	112	107	101	96	91	87	82	78	70
105	210	199	188	178	169	160	151	143	135	128	121	115	109	103	97	92	87	83	78	75
100	220	208	196	185	175	165	156	147	139	131	124	117	110	104	98	93	88	83	78	80
95	235	221	208	196	184	173	163	153	144	135	127	120	113	106	100	94	88	83	78	85
90	250	234	220	206	193	181	170	159	149	140	131	123	115	108	101	95	89	83	78	90
85	265	247	230	214	200	186	173	159	148	138	127	118	111	103	95	91	85	81	78	95
80	290	268	248	225	205	188	174	162	151	141	132	123	116	108	101	95	88	83	78	100
75	320	294	269	245	225	207	190	175	159	147	135	123	113	104	95	89	85	81	78	105
70	370	336	302	275	249	225	205	185	168	152	138	125	114	104	96	91	87	83	78	110
65	425	376	332	295	262	233	205	183	162	145	131	118	109	100	94	87	84	81	78	115
60	500	430	361	309	269	238	214	189	168	148	132	119	109	100	95	89	85	81	78	120
55	600	502	406	337	284	251	224	204	182	162	145	131	117	106	98	91	85	81	78	125
50	745	603	479	387	320	269	230	202	180	163	147	132	118	107	98	91	86	81	78	130
45	945	714	552	427	351	288	241	209	185	163	150	133	119	106	98	91	85	81	78	135
40	1185	871	655	507	406	324	263	216	190	166	148	132	118	107	100	93	88	83	78	140
35	1595	1200	910	676	504	380	288	219	178	152	136	120	109	100	94	88	85	80	78	145
30	2330	1710	1230	881	635	457	331	240	191	164	143	126	112	101	95	89	84	81	78	150
25	3700	2190	1445	1050	740	500	390	260	195	170	145	131	114	103	96	89	86	81	78	155
20	6100	2880	1700	1260	850	600	430	280	200	175	148	132	115	102	96	89	86	81	78	160
15	12000	3800	2190	1580	950	710	470	295	210	180	151	134	115	102	96	89	86	81	78	165
10	29000	5890	2950	2000	1100	810	520	320	230	190	158	137	115	101	96	89	86	81	78	170
5	140000	10700	4470	2880	1350	980	575	350	250	205	166	143	116	100	96	89	86	81	78	175
0	7.59(8)	89100	15800	6170	1820	1020	630	400	320	250	200	169	119	100	96	89	86	81	78	180

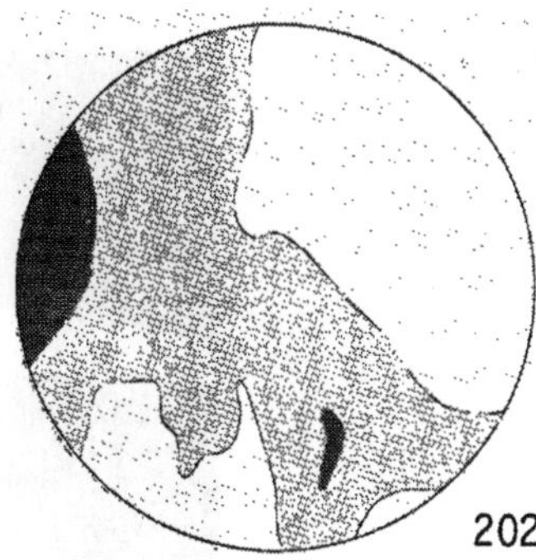

CHAPTER 4

2025

THE NIGHT AIRGLOW OR NIGHTGLOW

Our wandering planet is sheathed in a nocturnal glow that usually goes unnoticed. The glow is due to a variety of physical causes and, although it is often referred to as the night airglow or nightglow, it is probably better to speak of a number of nightglows, not only because of their numerous physical origins but also because of the multiplicity of their temporal and geographical changes.

Physical causes of the glow include chemical reactions of the neutral constituents of the upper atmosphere resulting in the emission of light; reactions involving ionized constituents also resulting in light emission; excitation of atmospheric constituents not in the Earth's shadow by sunlight; and excitation by incoming energetic charged particles guided along geomagnetic lines of force. The first two sources of the glow are omnipresent and contribute to the light of the night sky for all Earth-bound observers. The third has been observed recently from rockets and satellites, but much of its radiation is in the UV and does not penetrate to the Earth's surface. The fourth results in the polar aurora, which tends to occur in zones some 20° or 25° from the geomagnetic poles; the occurrence of the polar aurora has a strong dependence on the sunspot cycle.

The nightglow is a constant Earth 'envelope' – every astronaut has seen it on every 'night' orbit of the Earth, as a luminous annulus following the curved Earth surface (Figure 4-1). An astronaut in orbit around the Earth is particularly well located to see the nightglow, since his tangential view through the emission layer increases the effective path length some 30-fold relative to a zenith observation on the Earth's surface. The correspondingly enhanced brightening of the emission layer near 100 km puts it well above visual threshold.

From the Earth's surface there is a zenith-to-horizon increase in the radiance of the nightglow (Appendix 4-A), which can be observed in spite of the photometric dilution of the effect by the other constituents of the LONS. As a matter of fact, *each*

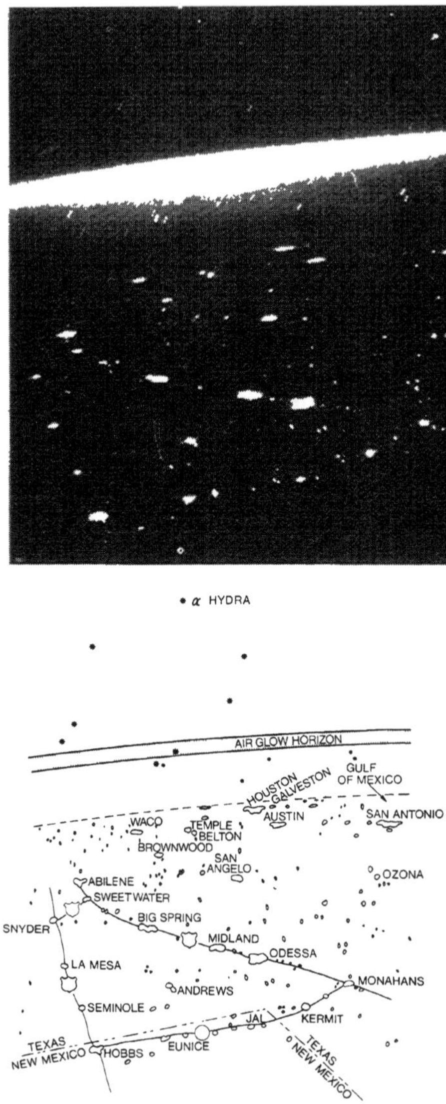

Fig. 4-1. Photograph and matching map of nightglow made during the Aerobee rocket flight of November 30, 1964. The nightglow is seen as a sort of annulus above the moonlit Earth over a large portion of Texas, with highways and towns identified on the map below (from Hennes and Dunkelman, 1966: 756–757).

major component of the LONS tends to dilute the contrasting effects of its competitors. It has been known since the star gaging of William Herschel that the region of the Milky Way is some 10 times as rich in stars as the sky well away from it (see Chapter 2). But, if the reader will position himself on a moonless night where he can look

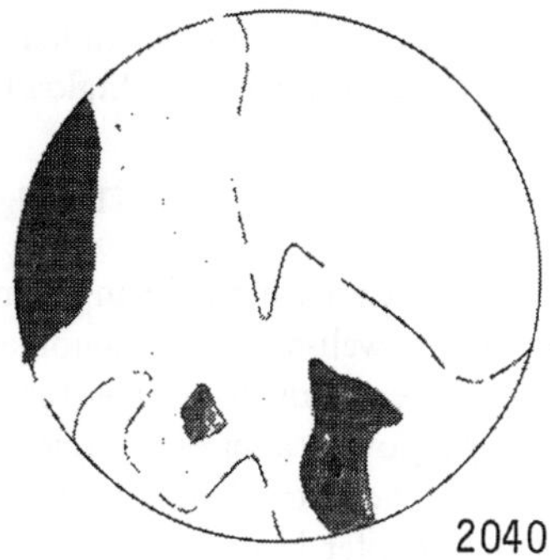

2040

at the sky*, he will quickly convince himself that the Milky Way – even in its brightest parts, such as the Sagittarius-Scorpius region – does not *look* 10 times as bright as the non-Milky Way sky. Obviously there is a dilution of the contrast by light sources of nonstellar origin. We have seen in Chapter 3 that a significant part of the photometric dilution is due to the presence of the zodiacal light, but this effect can be minimized if the observation is made during the middle of the night, when neither the bright evening nor the bright morning zodiacal light is a marked feature of the sky. If he will then make a vertical scan from zenith to horizon – avoiding the Milky Way as much as possible, and also avoiding the general east and west horizons because of possible residual zodiacal light – he can usually note that the night sky increases perceptibly in brightness toward the horizon as a result of the nightglow.

An early application of the zenith-to-horizon scan method was made by Yntema (1909), who dubbed the terrestrial component of the LONS the 'earthlight'. This component was investigated by Lord Rayleigh IV – the 'airglow Rayleigh' (see Plate V) – in the 1920s in early efforts to distinguish among various sources of the LONS. Having determined that this faint component was essentially unpolarized, he was able to eliminate the polarized zodiacal light as its major source. He then initiated photometric studies over the spectrum, obtaining evidence that led him to conclude that the faint light was separate from the polar aurora, and he named it the 'nonpolar aurora'.

The term 'airglow' was suggested by Otto Struve** to C. T. Elvey, who introduced it into the literature as '... the light emitted by the upper atmosphere other than that which is known as the polar aurora'† (1950). The generic term 'airglow' is often

* We assume that our observer has become reasonably dark-adapted; he might carry with him a black card to hold up against the sky – the contrast in brightness between the card and the sky is convincing evidence that the night sky is truly 'bright'.

** Though not an active worker in airglow research, Struve, as director of the Yerkes/McDonald Observatories, encouraged some of his junior associates (including F. E. Roach) to include geophysical airglow studies in general investigations of the light of the night sky.

† Elvey's definition carries a reasonable meaning to active workers in the field, even though it is often not possible to be definitive about the presence or absence of a polar aurora at any particular time or place. A random observation must of necessity await the *ex post facto* determination of auroral activity before it is known whether airglow has been observed.

particularized for specific studies – for example, day airglow (or dayglow), twilight airglow, night airglow (nightglow).

4.1. The Nightglow – Static or Dynamic?

The comparative difficulty in 'seeing' the nightglow has led to its being described as a quiescent, well-behaved phenomenon. Recent photometric studies indicate otherwise, however, and to counter that static view, we illustrate the dynamics of the tropical nightglow by a series of circle maps of 6300 Å centered on the Haleakala station in Hawaii. The 18 maps have been located on the upper right corners of the recto pages of this chapter in such a way that, by flipping the pages, the reader can produce a cinema effect as the localized patches of radiation wax and wane. In this time-lapse representation, the natural phenomenon is speeded up about 1000-fold – that is, an entire night of 6300 Å patterns is scanned in a few seconds. If the tropical enhancements, typically lasting an hour, were visible in real time, they would probably seem a bit sluggish; but to the photometric observer of the nightglow, they are quite dramatic, often requiring frequent changes of the sensitivity setting of the photometer to keep the deflections on scale.

As you will note on examining the corners individually, the time-lapse intervals extend over a period of 5 hr 30 min, during which more than a single waxing-waning cycle is revealed. If this phenomenon appeared in 'living color' in the night sky, it would be only slightly less spectacular than the polar aurora. As it actually occurs in the red below visual threshold, we regretfully can 'see' it only photometrically. Nevertheless, as the cinema effect demonstrates, it is a decidedly dynamic phenomenon.

4.2. Sources of Nightglow

The three principal components of the LONS (starlight, zodiacal light, and nightglow) – none of which is of overwhelming relative brightness – must be disentangled by successive approximations. The discrimination is aided by the fact that the nightglow, in contrast to the zodiacal light and the integrated starlight, includes many spectroscopically discrete emissions that can be resolved by spectroscopic methods – either a spectrograph of modest dispersion or filters with transmission bands centered on the wavelength of a particular radiation. For example, it was early established that the green line due to atomic oxygen (O I) at 5577 Å, a prominent auroral emission, (1) is always present in the night-sky light, (2) tends to increase in brightness toward the horizon, and (3) is not appreciably influenced by geomagnetic activity. Rayleigh IV made the first reliable measurement of the absolute brightness of this line and defined the photometric unit for expressing that measurement quantitatively. The unit, officially named the *rayleigh* (R)* in his honor in 1956, is in common use in airglow research today.

Both the nightglow and the aurora contribute to the LONS of an Earth-surface

* The unit is defined in Appendix 2-B.

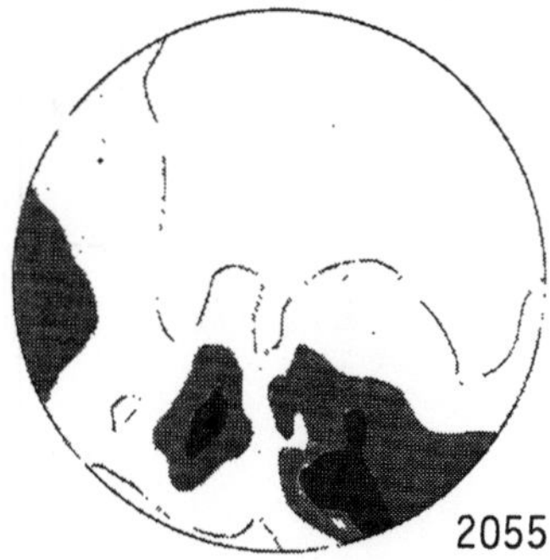

observer. The excitation of an aurora is the result of the transfer of some of the kinetic energy of the motions of impinging particles to the resident atoms or molecules in the upper-atmosphere zone. All other nocturnal upper-atmosphere radiations are nightglow. Because of its concentration in reasonably well-defined geographic zones*, we propose to treat the polar aurora as a perturbation – albeit a dramatic one – on the general planetary nightglow situation and discuss it separately at the end of this chapter.

There has been some confusion over the semantics of the expressions *nightglow* and *aurora* resulting in part from the contrast between the apparent dynamics of the two phenomena and in part from the original definition of the (night) airglow as upper atmosphere emissions of nonauroral origin. To some extent, a definitive appellation for a particular upper-atmosphere phenomenon producing radiation is not possible and, in such cases, it is probably wise to describe the phenomenon and search for its physical origins without undue preoccupation with names. A case in point is the M/SAR arc, which bears the earmark of an aurora in its undoubted orientation along geomagnetic parallels but is atypical in its very low excitation level, as discussed later in this chapter.

4.3. Photochemical Reactions in the Upper Atmosphere

In Figure 4-2 we show a schematic representation of the two-way interactions between light and chemical (or physical) changes in the upper atmosphere. Radiation (sunlight) produces changes in the atmospheric components by dissociating the molecular species into their atomic constituents and by removing electrons from neutral atoms or molecules and producing a plasma of charged particles intermingled with the still-predominant neutral particles. The products of dissociation or ionization interact to produce radiation (nightglow) as they recombine in the course of their small-scale (kinetic) or large-scale (geostrophic) motions.

In Tables 4-I and 4-II we list the principal nightglow radiations in decreasing order of absolute intensity. Table 4-I includes emissions resulting from the photo-

* Although an aurora may occur anywhere above the Earth's surface, it is statistically concentrated in the so-called auroral zones, small circles about 23 deg from the geomagnetic poles.

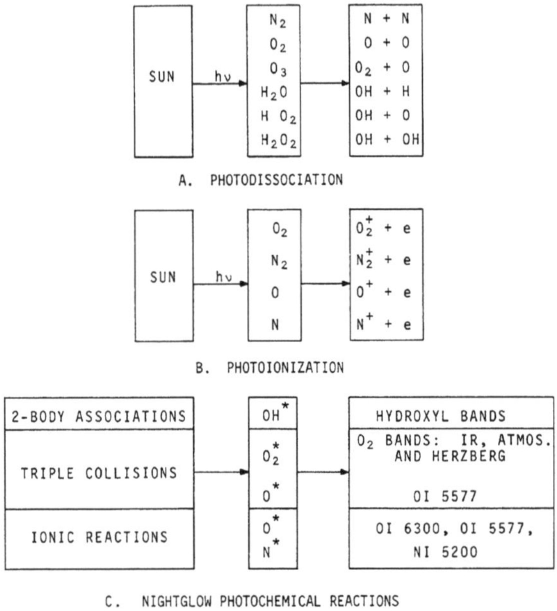

Fig. 4-2. Photoreactions in the upper atmosphere.

chemical association of neutral-atmosphere components, Table 4-II from reactions in the ionosphere involving ionized components. These groupings bring out the fact that the neutral reactions are effective in the 90 to 100 km region, the ionic reactions between 250 and 300 km.

4.3.1. Photochemical association

The 90 to 100 km region has often been called the 'chemical kitchen' because of the large number and the complexity of the chemical reactions that occur there. The complexity arises because the composition of the atmosphere in this region is significantly altered by the UV solar radiation. The total number density is shown in Figure 4-3. The principal effect is the dissociation of the O_2 molecule into two O atoms by solar radiation of wavelength shorter than 1750 Å, corresponding to the dissociation potential (7.049 eV) of the O_2 molecule (Figure 4-4). Superimposed on this photo effect are gravitational and dynamical influences such as the differential diffusion of components of different molecular weights and the somewhat compensating homogenization resulting from convection and turbulence.

In Figure 4-5 we show a model of the distribution with height in the 65 to 105 km region of some of the atmospheric components of interest in interpreting the nightglow. Atomic O (at midnight) peaks at about 92 km. Molecular and atomic oxygen are, according to this model, of equal concentration at 100 km. Molecular nitrogen (N_2) is not significantly dissociated at this altitude – its plot parallels the 'total' curve over this domain.

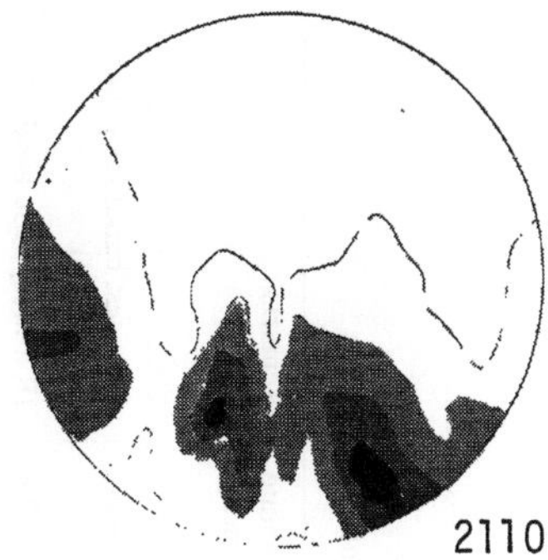

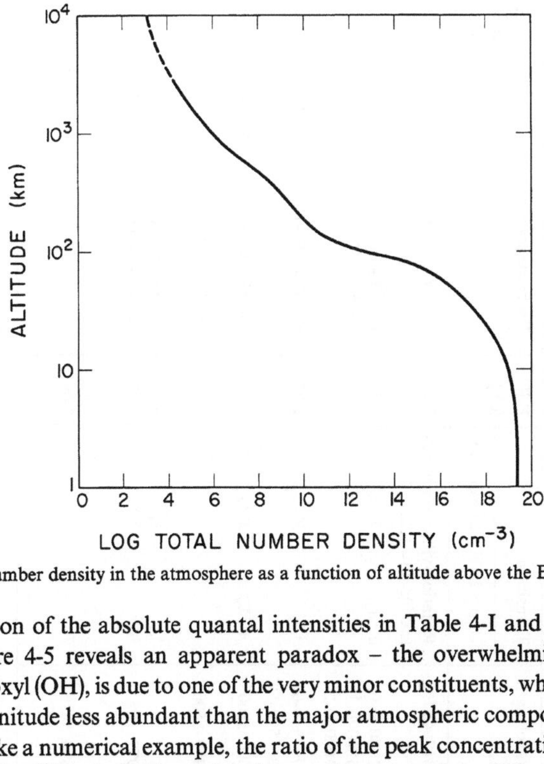

Fig. 4-3. Number density in the atmosphere as a function of altitude above the Earth's surface.

A comparison of the absolute quantal intensities in Table 4-I and the concentrations in Figure 4-5 reveals an apparent paradox – the overwhelmingly strongest emitter, hydroxyl (OH), is due to one of the very minor constituents, which is five or six orders of magnitude less abundant than the major atmospheric components – O, O_2, and N_2. To take a numerical example, the ratio of the peak concentrations of OH and O is about $10^7/10^{12} = 10^{-5}$. The ratio of quantal output of the OH system of bands to that of the O I line at 5577 Å is $4\,500\,000/250 = 18\,000$. Thus, if the emission is due to some mechanism concerned with the direct excitation of the emitting species, then the relative excitation efficiencies are

$$\mathrm{OH/O} = 18\,000/10^{-5} = 1.8 \times 10^9 .$$

TABLE 4-I
Nightglow radiations due to excitation by chemical association ordered by emission intensity

Emitter	Wavelength (Å)	Transition		E.P. (eV)	Probable reaction	Emission height (km)	Absolute intensity (R)	Remarks
		Upper level	Lower level					
Hydroxyl (OH)	3817–44702	Rotation-vibration bands ($v \leqslant 9$); $\Delta v = 1, 2, \ldots 9$		3.23 ($v = 9$)	$H + O_3 \rightarrow OH + O_2 + 3.34$ eV	90	4500000	Strongest bands are in near IR (Figure 4-4)
Molecular oxygen (O_2)	12700 (0-0) 15800 (0-1) 19000 (0-2)	$a\,^1\Delta_g$	$X\,^3\Sigma_g^-$	0.98	$O + O + M \rightarrow O_2\,(a\,^1\Delta_g) + M$ $k = 5.4 \times 10^{-34}$	90	80000	IR atmospheric bands
	7619 (0-0) 8645 (0-1)	$^1\Sigma_g^+$	$X\,^3\Sigma_g^-$	1.60	$O + O + M \rightarrow O_2 + M + 5.17$ eV	~80	6000	Atmospheric bands
	2600–3800	$A\,^3\Sigma_u^+$	$X\,^3\Sigma_g^-$	4.3	$O + O + M \rightarrow O_2 + M + 5.17$ eV	90	600	Herzberg bands
NO_2	5000–6500				$O + NO \rightarrow NO_2 + 3.1$ eV	90	250	Nightglow continuum
Atomic oxygen (O)	5577 2972	^1S ^1S	^1D ^3P	4.17 4.17	$O + O + O \rightarrow O_2 + O + 5.17$ eV $O + O + O \rightarrow O_2 + O + 5.17$ eV	90 90	250	 Observable only from rockets or satellites above ozone layer
Atomic sodium (Na)	5890, 5896	P	^2S	2.09		~92	50	Strong seasonal variation

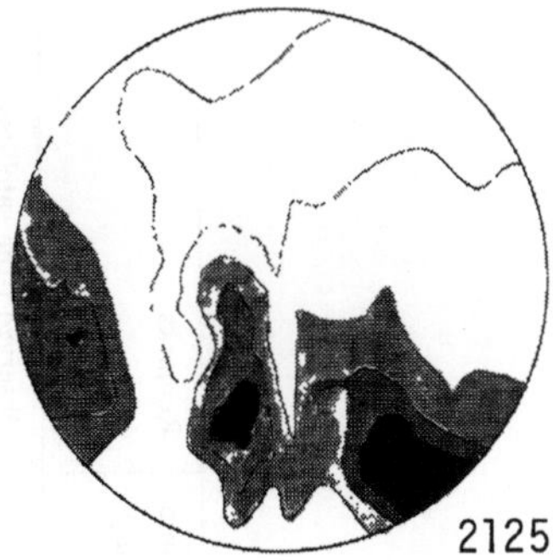

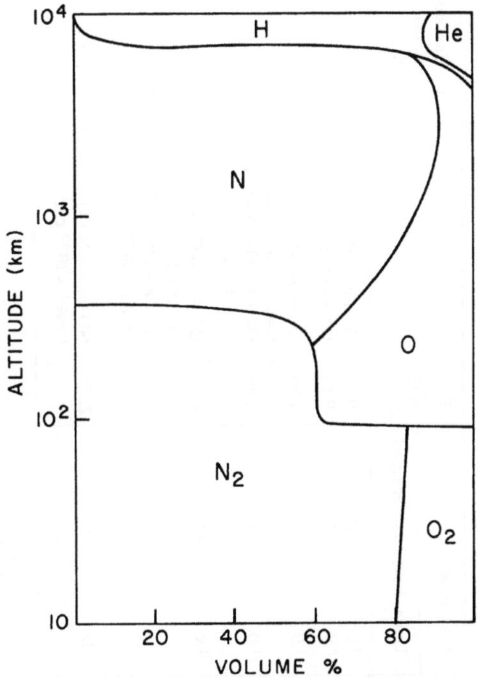

Fig. 4-4. Schematic representation of changes in chemical composition of the atmosphere with altitude. Note the abrupt change near 100 km, where O_2 molecules are significantly dissociated into O atoms.

It is obvious that we must anticipate entirely different types of excitation mechanisms for the two cases.

4.3.1.1. *Hydroxyl (OH) Emissions*

We have commented on the quantal output of OH emissions relative to that of the

TABLE 4-II

Nightglow radiations due to excitation by ionic reactions

Emitter	Wavelength (Å)	Transition		E.P. (eV)	Probable reaction	Emission height (km)	Absolute intensity (R)	Remarks
		Upper Level	Lower Level					
Atomic oxygen (O)	1304 1356	^3S ^5S	^3P ^3P	9.52 9.14	$O^+ + e \rightarrow O^* + h\nu$ $O^* \rightarrow O + h\nu$	250	150	Observed from satellites
	6300	^1D	^3P	1.96	$O^+ + O_2 \rightarrow O_2^+ + O$ $k = 1.5 \times 10^{-11}$ cm^3 s^{-1}	300	100	Sporadic enhancements in the tropics associated with ionospheric disturbances
	6364	^1D	^3P	1.96	$O_2^+ + e \rightarrow O + O + 6.96$ eV $k = 1.9 \times 10^{-7}$ cm^3 s^{-1}			
	5577	^1S	^1D	4.17	$O^+ + O_2 \rightarrow O_2^+ + O$ $k = 1.5 \times 10^{-11}$ cm^3 s^{-1} $O_2^+ + e \rightarrow O + O + 6.96$ eV $k = 2.1 \times 10^{-8}$ cm^3 s^{-1}	300	20	This high-atmospheric component of 5577 is observed chiefly in the tropics during enhancement of O I 6300, 6364
Atomic nitrogen (N)	5198 5201	$^2D^0$	$^4S^0$	2.37	$NO^+ + e \rightarrow N + O + 2.74$ eV $k = 4.1 \times 10^{-7}$ cm^3 s^{-1}	258	1	

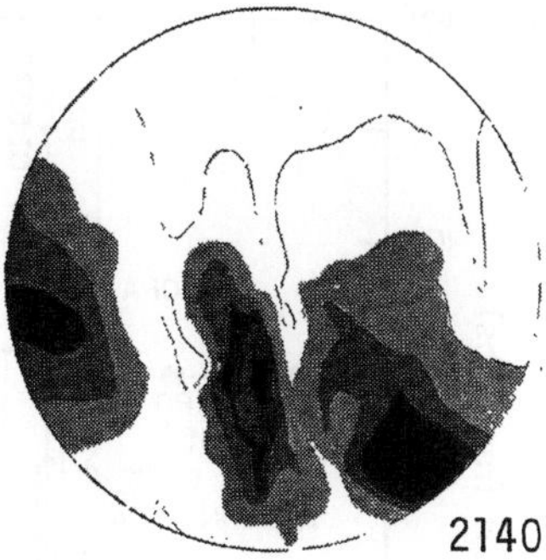

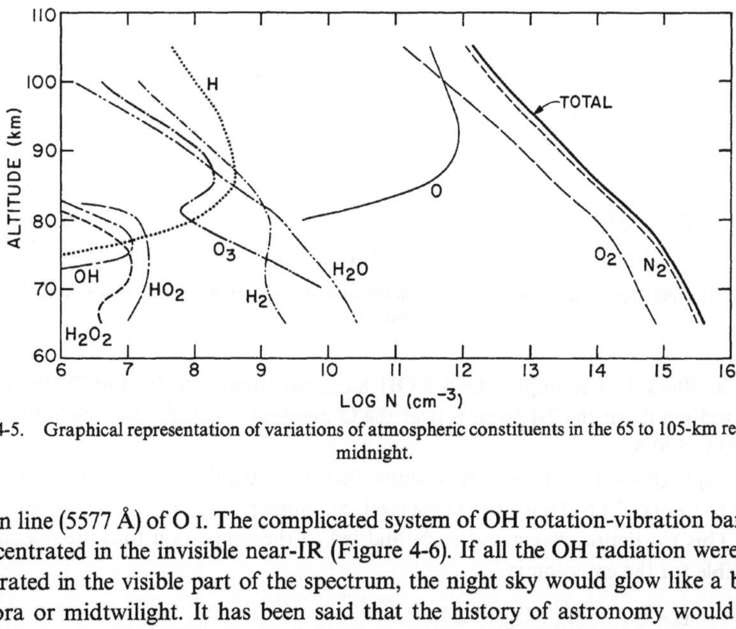

Fig. 4-5. Graphical representation of variations of atmospheric constituents in the 65 to 105-km region at midnight.

green line (5577 Å) of O I. The complicated system of OH rotation-vibration bands is concentrated in the invisible near-IR (Figure 4-6). If all the OH radiation were concentrated in the visible part of the spectrum, the night sky would glow like a bright aurora or midtwilight. It has been said that the history of astronomy would have been seriously affected by such a contingency, since only the brightest stars and planets would have been visible. The Milky Way, lost in the competing glow, would have been discovered only as a small photometric 'noise' fighting for recognition with respect to a very large OH 'signal'.

The OH system of bands is the result of transitions among the nine lowest levels, v, of the lowest electronic state ($^2\Pi$) of the radical. No selection rules apply and all combinations of Δv are permitted – 45 bands in all. Each band has a complicated structure (Chamberlain, 1961 : 368, table 9.1; 556, table 13.1), such that the individual lines spread out over some scores of Ångstroms. It is difficult to avoid some contamination by OH radiations in the case of observations of other nightglow emissions

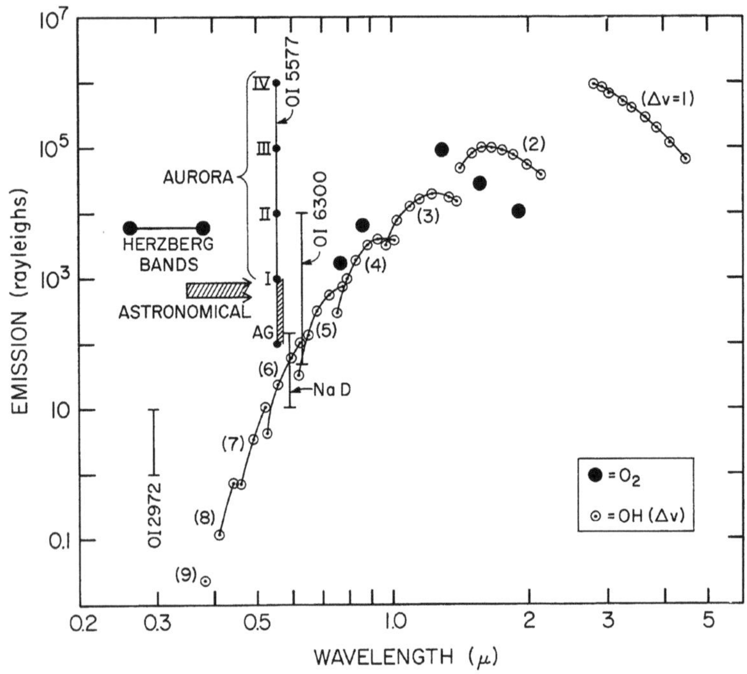

Fig. 4-6. Principal upper-atmosphere emissions on emission vs. wavelength plot. Compare with Tables 4-I and 4-II.

with wide filters. For example, the 8-2 OH band surrounds the D_1 and D_2 lines of atomic sodium (Na); the 7-1 band is near the O nightglow 5577 Å; and the 9-3 band is near O I 6300 Å.

In the nightglow OH a prime observational fact is that bands with upper vibrational level up to $v=9$ (3.23 eV) have been observed, but none from level $v=10$ (3.49 eV) or higher. This restriction was noted early and led to the suggestion that the reaction responsible for the emission is

$$H + O_3 \rightarrow OH + O_2 + 3.34\ \text{eV},$$

which is seen to be sufficiently exothermic to excite $v'=9$ but not $v'=10$. It remains to inquire whether this reaction between two very minor atmospheric constituents is sufficiently rapid to produce the observed quantal output. The rate coefficient,* k,

* The reaction rate constant k arises from the experimental law that the rate of reaction is proportional to the product of the concentrations of the reacting substances. k is the constant of proportionality. Thus, given a reaction $A+B+C=ABC$, the reaction rate is often written as a rate of loss, for example A, as $d[A]/dt=k[A]\,[B]\,[C]$. This definition applies to all types of reactions, whether the left side consists of one substance two, three, or more. The dimensions of k are different for different orders of reaction. Thus s^{-1} for first order, $cm^3\ s^{-1}$ for second order, $cm^6\ s^{-1}$ for third order.

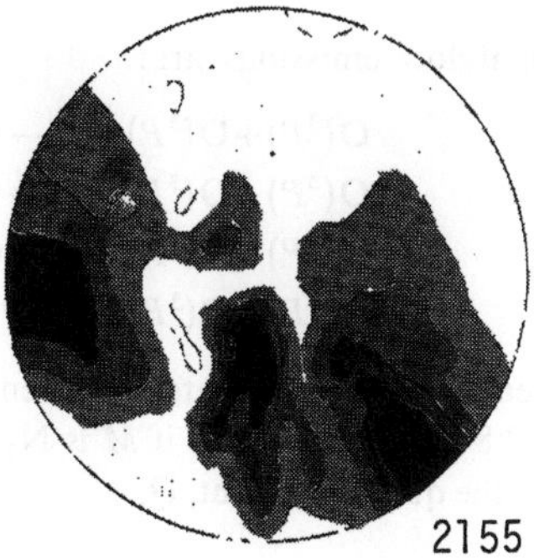

2155

of the reaction has been measured in the laboratory to be

$$k = 1.5 \times 10^{-12} \sqrt{T}\ \mathrm{cm^3\ s^{-1}} = 2.1 \times 10^{-11}\ \mathrm{cm^3\ s^{-1}}$$

for $T = 200\,\mathrm{K}$. At a height of 90 km, the predicted quantal output using this rate coefficient is

$$Q = [\mathrm{H}]\,[\mathrm{O_3}]\ 2.1 \times 10^{-11} = 10^8 \times 10^8 \times 2.1 \times 10^{-11} = 2 \times 10^5$$

quanta $\mathrm{cm^{-3}\ s^{-1}}$. At 85 km the prediction is 1.6×10^6 quanta $\mathrm{cm^{-3}\ s^{-1}}$. For an effective layer thickness of 10 km (10^6 cm) the columnar emission is 1.6×10^{12} quanta, or 1.6×10^6 R. Thus the general order of the prediction is in agreement with the observations.

4.3.1.2. *Molecular and Atomic Oxygen Emissions*

4.3.1.2.1. *Role of triple collisions in nightglow emissions.* We treat the three O_2 bands (Table 4-I) and the O I lines 5577 Å and 2972 Å together because the present evidence suggests a single basic excitation mechanism. In 1931, Chapman called attention to the reservoir of energy stored in the upper atmosphere from the recombining of atomic into molecular oxygen by way of a triple collision in which the energy of recombination (5.17 eV) is available either as kinetic energy of one of the reactants (thus raising the temperature) or as potential energy to excite one of the reactants to a state from which emission occurs. Triple collisions are infrequent as compared with double collisions, but have the property of using some of the excess energy (5.17 eV in the case under discussion) for excitation even when energy resonance is not involved.* The general reaction proposed by Chapman is

$$\mathrm{O} + \mathrm{O} + M \rightarrow \mathrm{O_2} + M + 5.17\ \mathrm{eV}, \tag{4.1}$$

where M is any third body, atomic or molecular. Particular reactions pertinent to

* Resonance, on the other hand, is favorable in the case of double-collision excitation.

nightglow emissions are:

$$O(^3P)+O(^3P)+M \rightarrow O_2(a\,^1\Delta_g)+M+4.19\text{ eV} \quad (4.2)$$
$$O(^3P)+O(^3P)+M \rightarrow O_2(^1\Sigma_g^+)+M+3.57\text{ eV} \quad (4.3)$$
$$O(^3P)+O(^3P)+M \rightarrow O_2(A\,^3\Sigma_u^+)+M+0.87\text{ eV} \quad (4.4)$$
$$O(^3P)+O(^3P)+O(^3P) \rightarrow O_2+O(^1S)+1.00\text{ eV}. \quad (4.5)$$

Reaction (4.2) has a rate coefficient of 5.4×10^{-34} cm^6 s^{-1}. If the concentration of O is 2.8×10^{11} cm^{-3} and if M is N_2 with a concentration of 7.6×10^{13} cm^{-3}, we have for the quantal output, Q,

$$Q = 6 \times 10^{36} \times 5.4 \times 10^{-34} = 3000 \text{ excitations cm}^{-3}\text{ s}^{-1},$$

or 3000 R for a column of 1 km (10^6 cm) thickness if each excitation results in an emission (no loss by collisional de-excitation). Thus the prediction is lower than observed for the IR atmospheric bands (80000 R). It has been suggested that the atmospheric model (Figure 4-5) provides too low a concentration of O in the 90 km region, possibly because of the downward movement of O during the night from the 100 km region. For example, if O has a concentration of 10^{12} cm^{-3}, reaction (4.2) predicts an emission (with no collisional de-excitation) of 40000 R for the IR atmospheric bands. However, the long mean lifetime of the upper state (3600 s) makes it subject to significant collisional de-excitation and imposes a severe restriction on the suggested reaction.

Reaction (4.5) has historical priority. Chapman introduced it as an excitation mechanism to produce O atoms in the ^1S state at a time when the 5577 Å ($^1D \leftarrow {}^1S$) line of O I was the best-known and most-extensively studied nightglow radiation. The reaction is the 'first born' of a large family. If the rate coefficient is the same as that for reaction (4.2), 5.4×10^{-34} cm^6 s^{-1}, then the peak quantal output is 540 quanta cm^{-3} s^{-1}, which corresponds to 540 R for an emitting layer of 10 km effective thickness (again assuming zero collisional de-excitation).

4.3.1.2.2. *Collisional de-excitation.* More than 40 reactions in the 100 km region have been listed among the allotropic forms of O and H: O_3, O_2, O, H_2, H, H_2O, HO_2, H_2O_2, OH. Of these reactions, some are directly concerned with the production of excited atoms or molecules that produce emissions. All the reactions are pertinent in an understanding of the total picture and, in particular, some of the reactions are competitive to emissions as de-excitation or quenching mechanisms.

Many of the nightglow emissions observed from the 100 km region are due to forbidden transitions. The upper excited atomic levels are metastable, having lifetimes long compared with the intercollision periods (see Table 4-I). The effect of collisional de-excitation during the long life of the typical metastable levels is now discussed for the case of the O I emissions 5577 Å ($^1D \leftarrow {}^1S$) and 6300 Å ($^3P \leftarrow {}^1D$). The energy-level diagram for the lowest states of O I (Figure 4-7) illustrates that the ^1S state has a mean life of 0.74 s and the ^1D state, 110 s. The collisional frequency at 90 km is

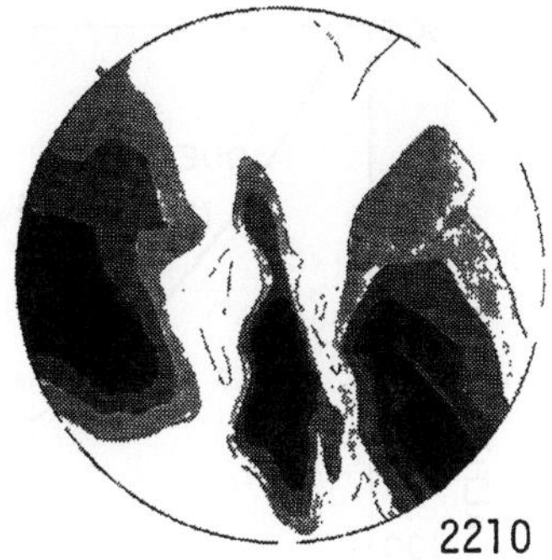

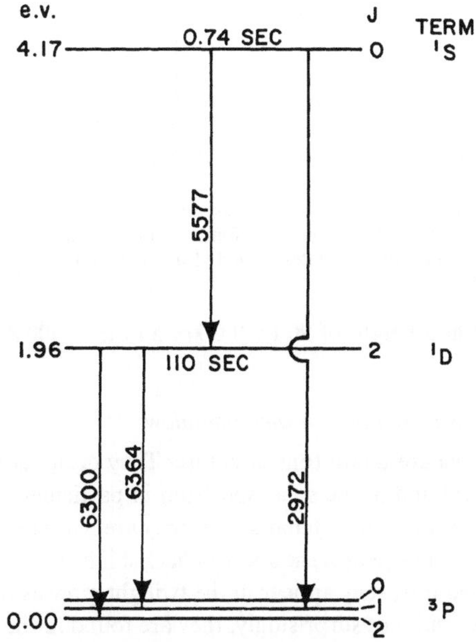

Fig. 4-7. Energy-level diagram of three lowest states of atomic O. All indicated radiations are due to transitions from long-lived (metastable) upper levels.

1.87×10^4 s^{-1}. Thus, during the time that an oxygen atom is procrastinating in the metastable ^1S state, it suffers 13 800 collisions and in the ^1D state, 2×10^6. Fortunately, the efficiency of de-excitation is low and only a small fraction of the collisions removes the atom from its excited state.

Figure 4-8 illustrates the change of de-excitation (quenching) with height based on an assumed quenching rate by collisions of 10^{-13} cm^3 s^{-1}. The deduced quenching is equal to the emission for the O I (^1S) state at a height of 100 km and for the O I (^1D) state at a height of 140 km. This is consistent with the facts that the 5577 Å emission

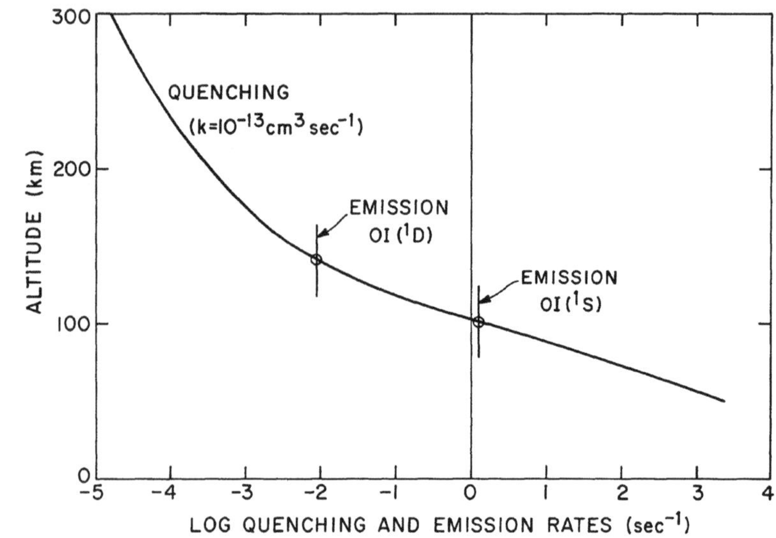

Fig. 4-8. Collisional de-excitation (quenching) rate in the upper atmosphere as a function of altitude for assumed de-excitation coefficient of 10^{-13} cm^3 s^{-1}. Emission rates (reciprocals of the mean lifetimes) for atomic O states ^1D and ^1S are indicated.

($^1D \leftarrow {}^1S$) is observed at a height of 90 to 100 km and the 6300 Å ($^3P \leftarrow {}^1D$) only at much greater heights.

4.3.1.3. *The Sodium (Na) D Lines in the Nightglow*

The yellow *D* lines of Na are ubiquitous in nature: They occur as absorption lines in stellar spectra in general and in the solar spectrum in particular; they show up as a component of the spectrum of interstellar space; they are scattered in the atmosphere from urban street lights; they plague the spectrochemist if he is not meticulously clean in his laboratory procedures; they appear in the twilight glow as the Sun illuminates the upper atmosphere; and, not surprisingly, they are found in the nightglow.

The very short lifetime of the upper atomic state, about 10^{-8} s, is one reason for the pervasive nature of the radiation. Such short residencies in the upper state mean that the $^2P^0$ Na atom at 90 km is able to radiate 5350 times between elastic collisions and about 16×10^6 times between quenching collisions. Even at sea-level pressures, there are about 50 atomic emissions per quenching collision.

The abundance of Na atoms in the upper atmosphere is about 100 cm^{-3}, too small to have been plotted on Figure 4-5. The low concentration is obviously compensated by the high transition rate in producing an observable radiation in the nightglow. An interesting feature of the Na nightglow is its seasonal variation. The radiance shows a pronounced winter (both hemispheres) maximum. In the local summer it is difficult to observe because of the low level of its quantal output.

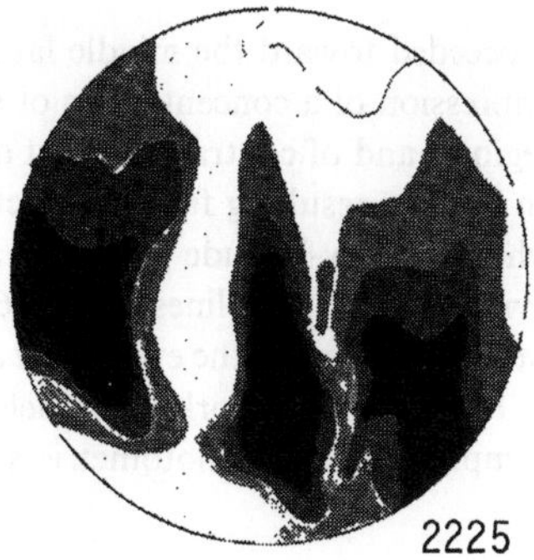

4.3.2. The nightglow and the ionosphere

Table 4-II brings together the facts concerning the nightglow radiation resulting from reactions involving ionized components of the upper atmosphere. The ionosphere is the consequence of the family of reactions indicated schematically in Figure 4-2b. For the details of the complex physical processes occurring in the ionosphere and the influence of the ionosphere on radio propagation, the reader is referred to standard works such as Rishbeth and Garriott (1969).

For the purposes of this book it is sufficient to point out two facts – first, the concentration of ionized components between 100 and 500 km is low relative to that of the permanent atmospheric components (Figure 4-9); and second, the reaction coefficients may be very high, especially when both reactants are electrically charged.

An astronaut in a polar-crossing orbit would be impressed with the dramatic drop in visual brightness of the upper atmosphere as he crossed the auroral zones and

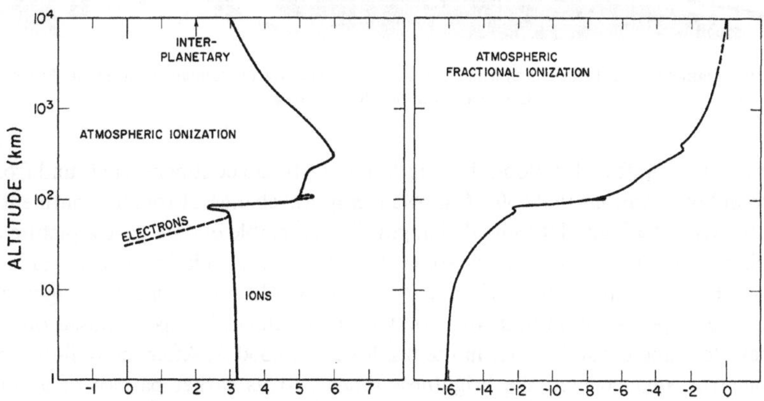

Fig. 4-9. Variation of nocturnal ionization in the atmosphere with altitude. Peak of about 10^6 electrons cm^{-3} represents a fractional ionization of only about 10^{-4}.

proceeded toward the middle latitudes and then on toward equatorial regions. The impression of a concentration of bright and dramatic optical displays in the auroral regions, and of contrasting faint and quiescent events in the lower latitudes is partly an illusion resulting from the fact that his eyes are not sensitive to the wavelengths where the low-latitude emissions are active. Especially interesting is the tropical activity of the O I red lines 6300/6364 Å, which we shall refer to generically as 6300 Å. At this wavelength the eye is only about 1/10 as sensitive as it is in the green.

The pioneering work in this field was done by D. Barbier (Figure 4-10), who in 1957 complemented his photometric studies at the Tamanrasset observing station in the

Fig. 4-10. Daniel Barbier (1908–1965) with his multi-color nightglow photometer at the Haute Provence Observatory in Southern France.

Sahara with aircraft flights along the length of the African continent. He found from his ground observations that 6300 Å showed temporal changes. From his continental aircraft sweeps he found latitudinal changes. The ensemble of data gave a picture of two photometrically active regions about 20 deg on either side of the geomagnetic equator. The resulting photometric activity is a many-fold enhancement in brightness lasting over a period of an hour or two. The enhancement is superimposed on the 'typical' slow nocturnal decrease in the brightness of 6300 Å. After the well-defined photometric enhancement, the brightness returns to its earlier pattern of a slow decline in brightness (Figure 4-11).

That the source of the photometric perturbation in the brightness of 6300 Å is a local enhancement in the concentration of the ionized components is suggested by the

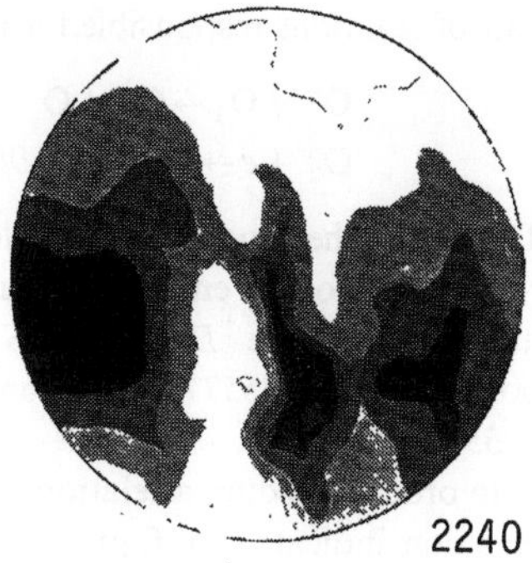

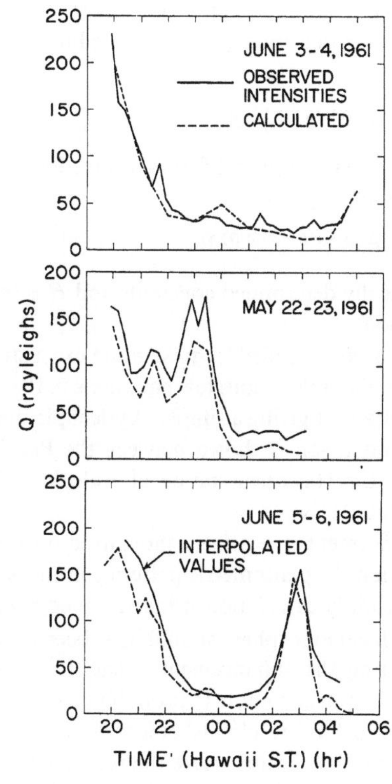

Fig. 4-11. Three examples of nocturnal variations of 6300 Å at tropical station Haleakala (Hawaii). Upper plot is typical for nights when there is a steady decline in brightness during the night. Middle and lower plots illustrate cases in which significant interruptions in the photometric pattern occur. Dashed curves are computed by Barbier's Equation (4.8), using ionospheric parameters (see text) for comparison with observations (solid curves).

pair of reactions responsible for the photon emission:

$$O^+ + O_2 \rightarrow O_2^+ + O \qquad (4.6)$$

$$O_2^+ + e \rightarrow O + O + 6.96\ \text{eV}. \qquad (4.7)$$

The rate of the process is controlled by the first reaction. The photochemical reaction responsible for the emission (Equation (4.7)) is sufficiently exothermic to excite the O atoms to both the ^1D and the ^1S states, which action is consistent with the fact that both 6300 Å and 5577 Å are found in the tropical enhancements in the ratio Q 6300/Q 5577≈5.

In order to deduce a relationship between ionospheric parameters and the quantal emission indicated by Equations (4.6) and (4.7), Barbier made use of the fact that ionospherists systematically characterize the state of the ionosphere by the use of two quantities that define the degree of ionization as well as the locus of its effective height: f_0F_2, the critical frequency of the F layer in the ionosphere, and $h'F$, the virtual height to the layer. The peak electron density n_0 (max) is related to the critical frequency by the relationship

$$n_0(\text{max}) = 1.24 \times 10^4\ (f_0F_2)^2 \text{ electrons cm}^{-3},$$

where f_0F_2 is in MHz. The semi-empirical formula introduced by Barbier gave the quantal output of the emissions, Q, by

$$Q = A + B(f_0F_2)^2\ e^{[-(h'F-200)/H]}\ \text{R}, \qquad (4.8)$$

where A and B are empirically determined constants and H is the atmospheric scale height in the emitting region.

Considerable effort has been expended to make more exact studies of the relationship between the ionosphere and the nightglow emissions in the tropics, but the Barbier formula has the advantage of giving a simple physical picture. Figure 4-11 shows three cases in which a comparison is shown between the Barbier formula and the observations of 6300 Å at the Hawaiian station at Haleakala (lat. N 20°43′; long. W 156°16′; geomagnetic lat. 21° N).

A typical tropical enhancement is relatively short lived during a given night, and at the end of the perturbation the photometric pattern continues as though there had been no interruption. A simple calculation using the prediction of Equation (4.7) indicates that the original local ionosphere should have been entirely 'burned out' by the loss of ionization during the enhancement. That the local ionosphere is *not* destroyed indicates that the phenomenon is probably associated with geostrophic movements of ionization, which bring 'clouds' of ionization to the particular tropical domain where the enhanced 6300 occurs. Here the reader might 'replay' the corner circle maps, which dramatically emphasize the enhancement.

An OGO (Orbiting Geophysical Observatory) satellite moving in an approximately polar orbit passes over the Earth's surface in about 90 min. Figure 4-12 shows a composite plot of the 6300 Å on a global basis, built up by a series of passes. This global representation is reminiscent of the pioneer work performed by Barbier in his flights

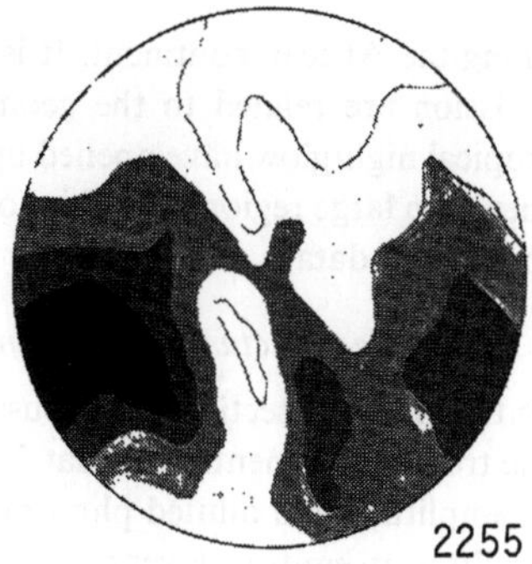

2255

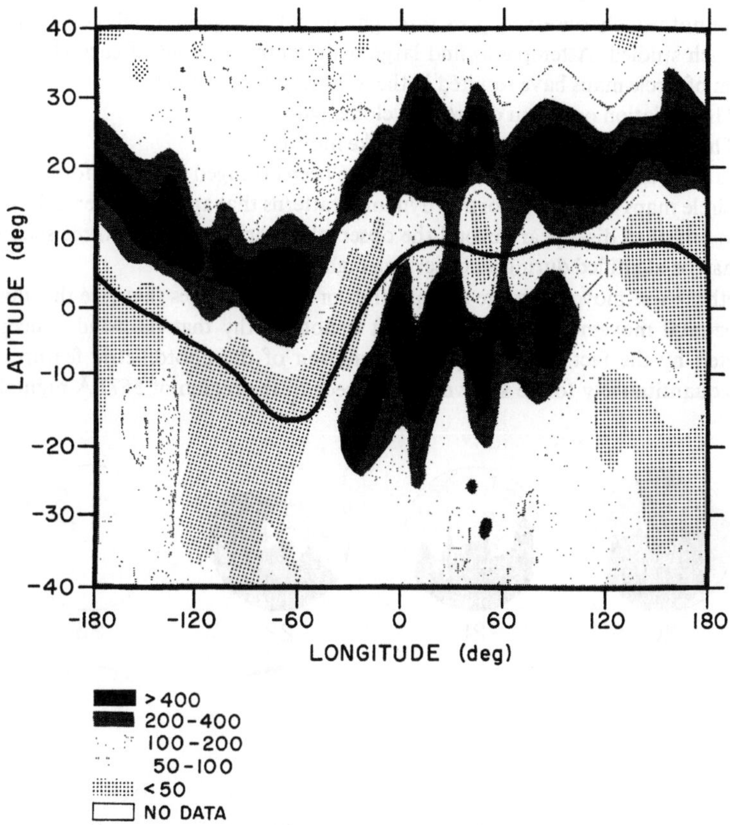

Fig. 4-12. Composite world-wide photometric map of 6300 Å made from a series of satellite passes (adapted from Reed *et al.*, 1973). The locus of the geomagnetic equator is indicated by a solid line, and the intensity scale, in R, is listed below.

along the African continent. It is apparent that the regions of concentrated 6300 Å emission are related to the geomagnetic equator. The photometric studies of the tropical nightglow have opened up a new domain of ionospheric exploration because they map large regions of the ionosphere in a synoptic way not possible from isolated ionosonde data.

4.3.2.1. *The Structure and Dynamics of* O I 5577 Å

In the previous section we discussed the temporal variations of 6300 Å nightglow in the tropics and mentioned that 5577 Å goes through a similar variation, not so large in amplitude and diluted photometrically. The major source of excitation of O I(^1S) results in an emitting layer of 5577 Å near the 100 km region, in contrast to the tropical ionospheric emission that occurs in the ionosphere above 200 km.

The 5577 Å emission, especially in midlatitudes where the ionospheric perturbations are not usually observed, goes through diurnal changes in brightness that have been much studied. Attempts to find large-scale representations of such changes in the form of circle maps have resulted in the following conclusions:

(1) The radiation occurs in discrete, cyclonic patterns.

(2) These patterns are in a sort of wind-like motion.

Maps (such as those in Figure 4-13) that have been arranged in cinema form (similar to the circle maps of 6300 Å) portray dynamic patterns that appear, traverse the circle of view, and then disappear beyond the observable horizon. Two or three such patterns may be identified during an observing night.

Whether these movements are actual winds or *in situ* changes that give the illusion of horizontal movement is not known. It is noteworthy that the wind concept is supported by the fact that the speeds of motion of the photometric features are similar quantitatively to the known wind speeds in the region of 5577 Å nightglow.

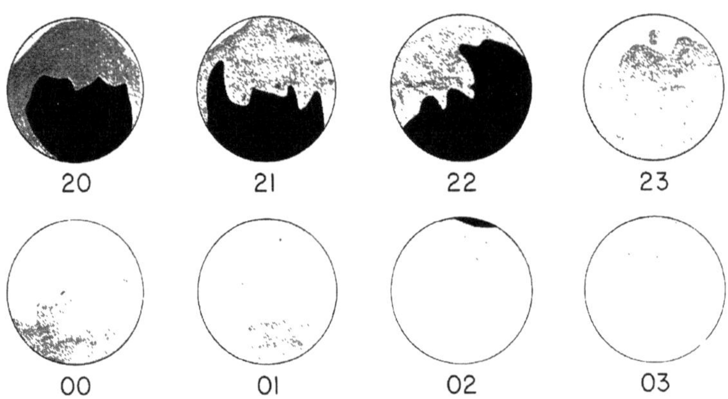

Fig. 4-13. Series of circle maps of 5577 Å from observations made at midlatitude station of Fritz Peak, Colorado. Circle diameter is about 1000 km. Numbers indicate hours in Mountain Standard Time (7 hr W).

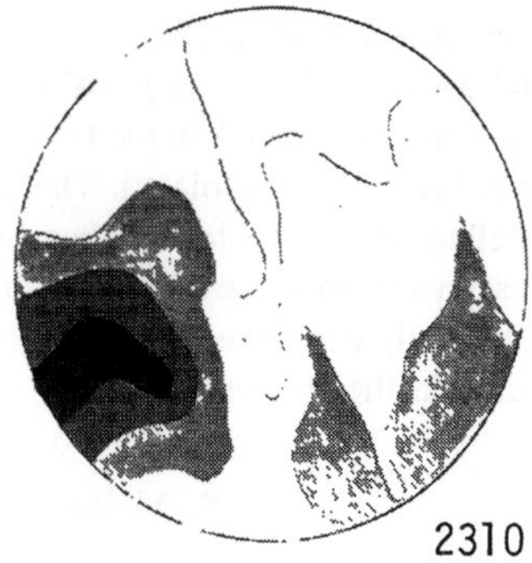
2310

4.3.3. THE NIGHTGLOW CONTINUUM

Whether there is a general spectoscopically structureless radiation from the upper atmosphere in the visual part of the spectrum is really a moot question. From the observational standpoint, the existence of such radiation is difficult to establish because spectroscopically it cannot be distinguished from the zodiacal light and the integrated starlight unless observing techniques could be devised of such high resolution that the individual Fraunhofer absorption lines of the astronomical components of the LONS could be resolved. The present evidence suggests that the so-called nightglow continuum is partly a real upper-atmosphere phenomenon and partly due to sunlight that has suffered multiple scattering in the atmosphere. Exact quantitative measurements from space will probably ultimately give the answer.

4.4. The Nightglow from Space: the Exosphere and Geocorona

We introduced this chapter by reference to the fact that the visual nightglow layer near 100 km, due chiefly to atomic O I 5577 Å, is a characteristic feature of the astronauts' look at the Earth's limb during satellite night. During the first full decade of the space age, the 1960s, there was, in parallel with the manned program, a strong program of observations from unmanned satellites and from rockets which yielded new information on the nightglow from topside.

Referring back to Figures 4-3 and 4-4, we note that the upper atmosphere from 1000 to 10 000 km above the Earth thins out to a density approaching that of interplanetary space. In this region, called the exosphere, kinetic collisions become very infrequent. The chemical composition changes to a H–He mixture and finally to an ionized plasma (Figure 4-9). At 1000 km, the frequency of collision is only about one in 10 min (contrasted to about 4000 s^{-1} at 100 km). A particle travels on the average about 500 km between collisions, and the statistical probability of the escape of a particle from the gravitational hold of the planet without any collisions with other particles becomes significant.

In this region and continuing out into a still rarer domain, a sort of geocorona has been observed, consisting of emissions from the H atom (the Balmer α line, 6563 Å, and the Lα, 1216 Å, and β, 1026 Å, lines) and from the He atom emissions at extreme wavelengths of 584 Å in the UV and 10830 Å in the IR. Also, an ionized He emission at 304 Å has been isolated. The UV emissions can be observed only from rocket or satellite above the lower absorbing atmosphere. Other emissions that have been observed from space are the permitted O I lines, 1304 Å and 1356 Å, which are ionospherically controlled and, like the 6300/6364 Å lines, are characteristically concentrated in the tropics.

4.5. Midlatitude Stable Auroral Arcs (M/SAR)

In 1958, one year after his report on the tropical enhancements of O I 6300 Å, Barbier, working at the midlatitude station of the Haute Provence Observatory, announced the discovery of an arc-like phenomenon that bears a superficial similarity to the tropical phenomenon discussed earlier (p. 64), in that the same radiation, 6300 Å, is involved. It differs in other respects, however, indicating that it must have a quite different physical origin. Barbier dubbed it *L'arc aurorale stable* and it has come to be known as a SAR arc, an anglicized acronym of Barbier's appellation. It is highly selective spectroscopically, favoring strongly 6300 Å, and it is a midlatitude phenomenon – hence it is also sometimes referred to as an M/SAR arc (monochromatic; midlatitude).

Its energetic selectivity, ^1D (1.96 eV) over ^1S (4.17 eV) (see Figure 4-8) is in marked contrast to the general spectroscopic nature of the polar aurora, which is definitely a high-energy phenomenon in terms of the excitation levels of the emissions. The M/SAR-arc is energetically different from the tropical case, in which the green line is enhanced along with the red.

M/SAR arcs are observed during times of geomagnetic activity and are clearly associated with auroral displays, though discretely to their equator side. They are oriented along geomagnetic parallels and hence must be associated with moving charged particles in the upper atmosphere. If the association is one of *direct* excitation of resident atoms by incoming charged particles, it becomes necessary to impose a strong upper limit on their energy such that the resident oxygen atoms are excited to the ^1D state (1.96 eV) but not to the ^1S state (4.17 eV). What seems more likely is that the excitation is caused *indirectly* by the raising of the temperature of the resident atoms.

In Figure 4-14 we show schematically the zones in which the three phenomena occur – the polar aurora, A; the M/SAR arcs, M; and the equatorial perturbations, E. The M zone is shown as separated from the A zone because, although the aurora may on rare occasions extend to equatorial regions, the M arcs have usually been observed discretely and equator-ward with respect to coincident polar auroras (Figure 4-15). The M/SAR arcs have their photometric center near 400 km (Figure 4-16) in contrast to the polar aurora, which extends down to the 100 km region. The geomagnetic field is obviously in control, as evidenced by the orientation of the photometric maxima along the geomagnetic-latitude circles.

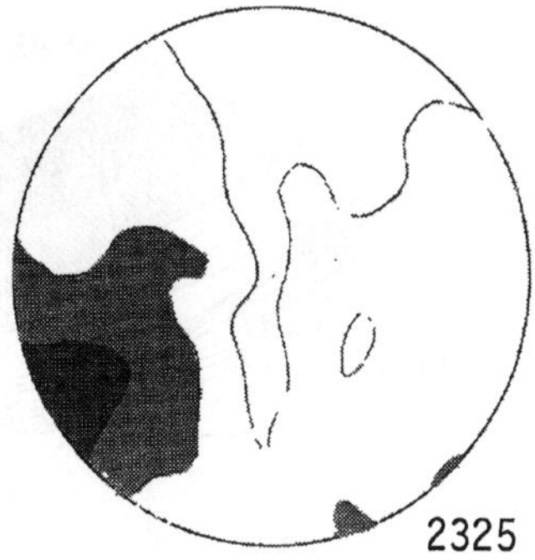

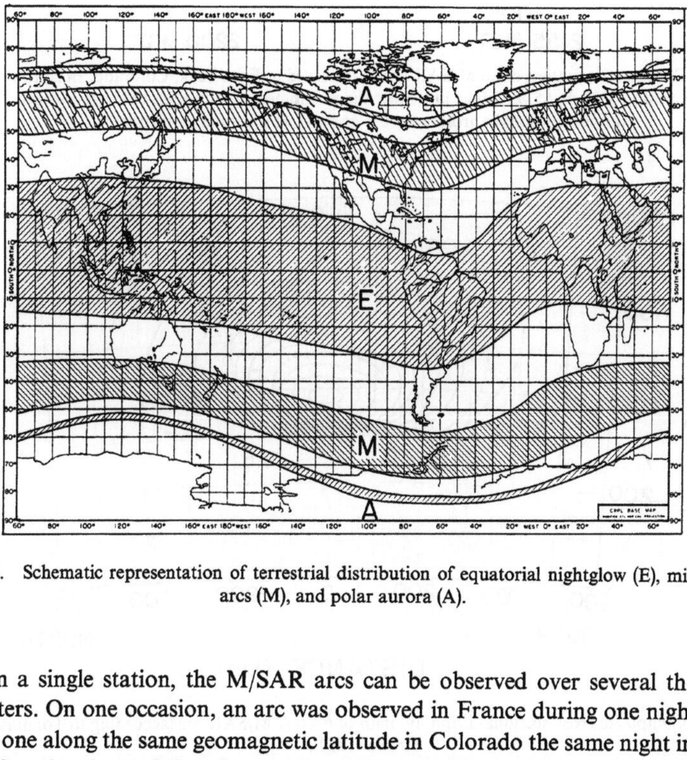

Fig. 4-14. Schematic representation of terrestrial distribution of equatorial nightglow (E), midlatitude arcs (M), and polar aurora (A).

From a single station, the M/SAR arcs can be observed over several thousand kilometers. On one occasion, an arc was observed in France during one night and a similar one along the same geomagnetic latitude in Colorado the same night immediately after the close of the observing in France, indicating that an excitation ring extended at least one third of the way around the globe. The absolute radiance of the arc has been measured as high as 8000 R. If this were at a wavelength near 5577 Å, the phenomenon would be a conversational spectacle, but the visual threshold of the eye in the red is about 20000 R (only 1000 in the green), so this dramatic display of

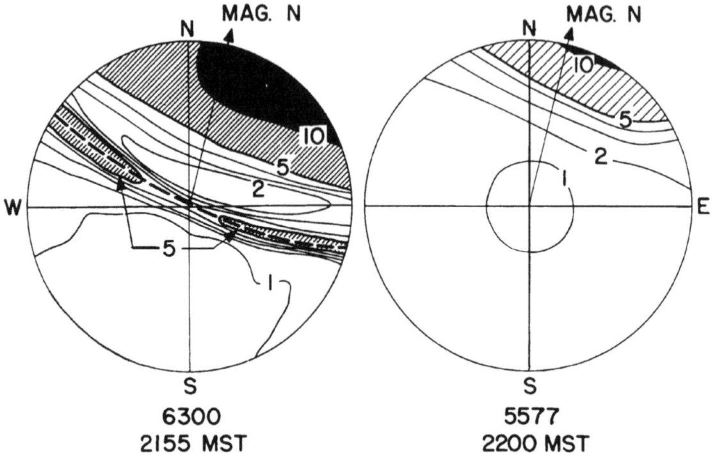

Fig. 4-15. Photometric pattern of an M/SAR arc as observed at Fritz Peak, Colorado. In contrast to the well-defined arc for 6300 Å, 5577 Å does not show the phenomenon, although both wavelengths display the conventional aurora near the horizon, toward magnetic north.

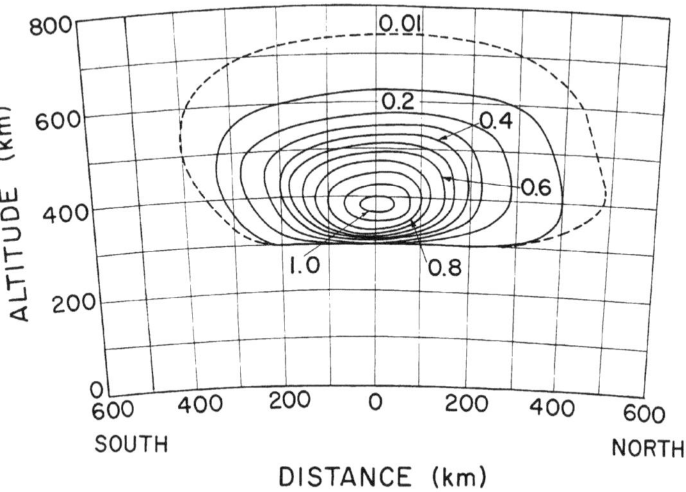

Fig. 4-16. Composite cross-sectional isophotal map of 6300 Å M/SAR arc (according to Tohmatsu and Roach, 1962).

nature must be 'seen' by instruments. If there were a small upward shift of 0.2 eV of the ^1D level, then the $^3P \leftarrow {}^1D$ radiation would be green, and the *M* arcs would be a source of great beauty. Of course, such a shift would move 5577 Å ($^1D \leftarrow {}^1S$) into the red, and the appearance of the polar aurora would be drastically changed.

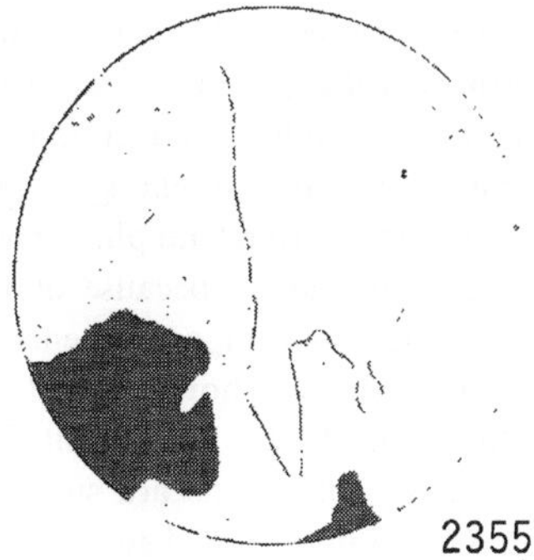

Since much of our research motivation has been initiated by the excitement arising from visual observations, we may conclude that a mere 0.2 eV in a single atomic-energy level has affected nightglow/aurora scientific history!

4.6. The Polar Aurora

The polar aurora differs from the nightglow in two observational features – one in degree and the other in fundamentals. First, the aurora is brighter than the nightglow,

Fig. 4-17. Photograph of polar aurora, looking toward the (geomagnetic) east. Reproduced from a color slide taken by Dr John Boyd of the Geophysical Institute, University of Alaska (with permission).

especially in the visible part of the spectrum (Figure 4-17). Second, the spectroscopic features of the aurora indicate a high level of excitation, being due to the transfer of energy from high-speed incoming particles, in contrast to nightglow excitation arising from the exchange of energy in photochemical reactions. The M/SAR arc is, in this respect, an intermediate phenomenon – it is a low-excitation case but is usually classified as an aurora because of its coincidence with auroras and its geomagnetic alignment. The comparative energetics of the upper-atmosphere emissions discussed in this chapter are shown in Table 4-III. The auroral brightnesses progress from International Brightness Coefficient (IBC) I, for which the vertical brightness of 5577 Å is 1000 R (1 kR) by decade steps to IBC IV, for which the 5577 Å brightness is 1000 kR. The summation in Table 4-III of *all* the polar aurora features is taken from Chamberlain (1961:197, table 5.5).

TABLE 4-III

Energetics of upper-atmosphere emissions

Phenomenon	Total photon emission (R)	Total energy emission[a] (ergs cm^{-2} s^{-1})
Nightglow		
OH	4500000	3.2
Other	88000	0.125
M/SAR Arc	8000	0.025
Polar Aurora		
IBC I	68200	0.18
II	682000	1.8
III	6820000	18.0
IV	68200000	180.0

[a] In a column.

The visual observer of the aurora often sees a dynamic picture of apparently rapid movements across the sky as the locus of the excitation changes. Colors are evident for auroras between IBC II and IBC III as the *color* threshold of the eye (about 20000 R) is exceeded. Mixes of O I 5577 Å, O I 6300 Å, Hα 6563 Å, and N_2 first positive bands contribute to a variety of color sensations. If 5577 Å predominates, the color is green; O I 6300 Å, when present, often produces a red upper overlay. The first positive bands of N_2 occur below the aurora proper, and when bright enough produce a red lower border. Even a pastel shade may result from a mixture of several radiations.

Polar auroras tend to occur in the so-called auroral zones. In addition to this spatial preference, there are well-defined temporal fluctuations. There are maxima of occurrence in the spring and autumn (Figure 4-18). Auroral activity goes through an 11 yr cyclical variation associated with the activity on the Sun as measured by sunspot numbers (Figure 4-19). The auroral maximum in general occurs about 2 yr after the sunspot maximum.

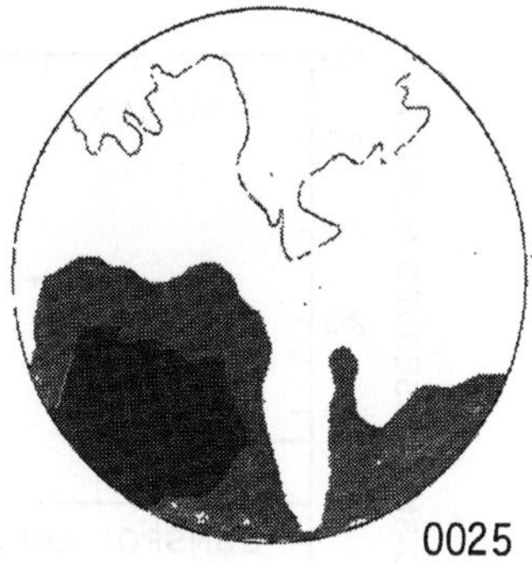

The excitation mechanism of a polar aurora is probably the direct excitation of the atmospheric constituents by charged particles proceeding along geomagnetic lines of force to the portion of the atmosphere where impacts are sufficiently numerous to decelerate the intruders, with multiple excitations resulting from the impacts. Bright auroras are of great beauty and are often dramatically variable in position, form, and color as the excitation conditions in their variability often give the impression that the auroral features are sweeping across the sky in a 'wind-like' motion of great speed.

When present, these brilliant displays are striking features of the sky. They preclude many astronomical studies for which a sky with a faint background (or foreground) is necessary, however. Therefore we take advantage of the fact that polar auroras tend to be concentrated in fairly restricted zones and occur somewhat sporadically

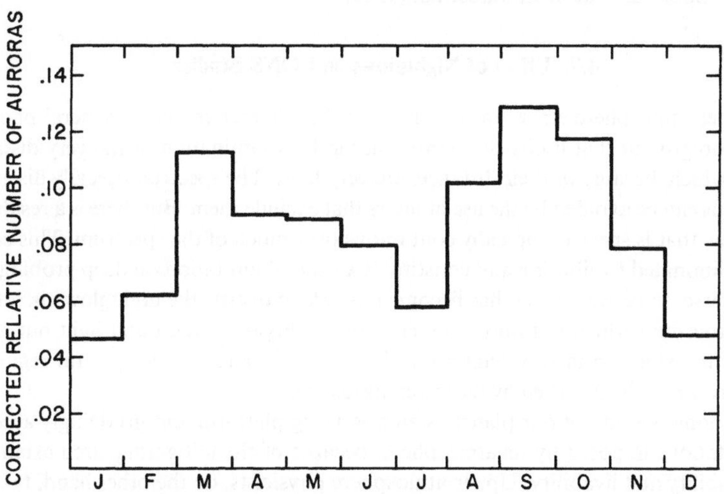

Fig. 4-18. Histogram showing seasonal variation of polar auroras as seen from Yerkes Observatory, corrected for cloudiness and number of dark hours each month (from Meinel *et al.*, 1954:411).

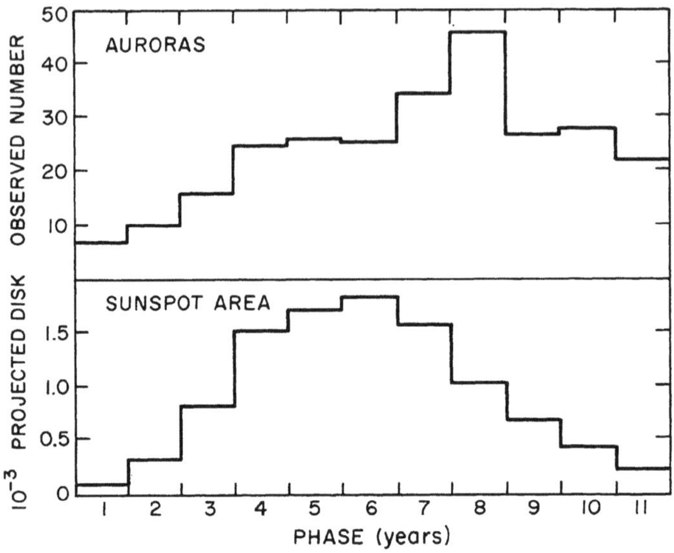

Fig. 4-19. Variation of auroral occurrence as seen from Yerkes Observatory during a sunspot cycle, based on a composite over several sunspot cycles (from Meinel *et al.*, 1954: 408). Zero phase refers to sunspot minimum.

in giving them only a comparatively brief treatment. The reader interested in an understanding of the complexity and the physical origin of the auroras should consult such books as that of Chamberlain (1961).

4.7. Effect of Nightglows on LONS Studies

The upper-atmosphere emissions discussed in this chapter present a general photometric foreground that limits the astronomer in his examination of the very distant objects which, because of their distance, are very faint. The spectroscopically discrete radiations can be avoided by the use of filters that exclude them. But there is a residual nightglow that is spectroscopically continuous over much of the spectrum. This cannot be eliminated by filtering and constitutes a general limitation on deep probing of the universe. In recent years it has become possible to overfly the nightglow via satellite, a procedure which not only eliminates the nightglow foreground light but also permits an extension in wavelength into both the UV and the IR spectral regions, which are seriously absorbed by the lower atmosphere.

Astronomers think of our planet as an observing platform and grudgingly accept the limitations imposed by its atmosphere, because of the mitigating circumstances of its stability and livability. Upper-atmosphere physicists, on the other hand, take a more parochial view of the situation, applauding the nightglow because of its contribution to an understanding of upper-atmosphere processes, accepting with some

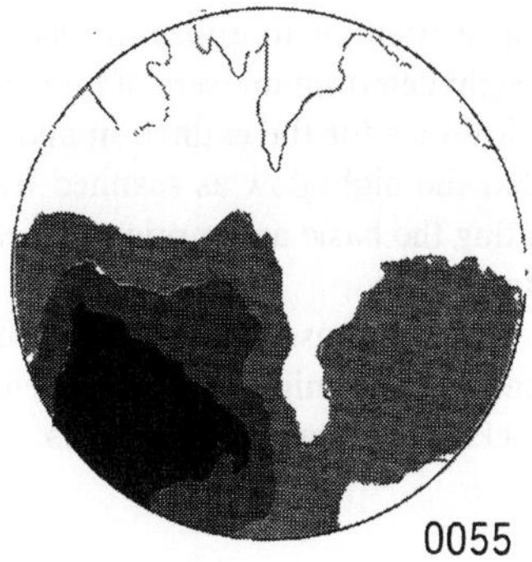

grace the technological difficulties imposed by the presence of light of astronomical origin in the night sky.

Appendix 4-A. The Change in Brightness of a Nightglow Layer with Zenith Distance

We have mentioned that a distinguishing feature of a nightglow layer is its increase in brightness from the observer's zenith toward his horizon. This change is due to the fact that an emission occurring in a uniform layer bounded by concentric arcs with respect to the center of the Earth will present an emission path length to an observer on the Earth's surface increasing with zenith angle, z, according to*

$$V=\left(\sqrt{1-\left(\frac{r}{r+h}\right)^2 \sin^2 z}\right)^{-1},$$

where r is the radius of the Earth and h is the height of the emitting layer above the Earth's surface. Some values of the function V are shown in Table 4-IV, with z and h as parameters.

TABLE 4-IV

The van Rhijn function, V, at different altitudes

Zenith angle (deg)	V at different altitudes			
	100 km	200	300	400
0	1.000	1.000	1.000	1.000
40	1.292	1.279	1.267	1.256
60	1.914	1.841	1.779	1.725
70	2.635	2.426	2.267	2.141
80	4.087	3.367	2.945	2.661
90	5.713	4.086	3.374	2.955

* The equation was first used by P. J. van Rhijn (1921) in an atmospheric connotation. The function, V, is often referred to as the van Rhijn function.

Early attempts to utilize the zenith-to-horizon increase of nightglow brightness as a height determinant were of minimal value because of: (1) difficulties in making proper allowance for the extinction and scattering of the lower atmosphere and (2) the fact that the nightglow as scanned by a given observer is intrinsically patchy, thus violating the basic assumption of a uniform emitting layer used in deriving the V equation.

For a discussion of these effects, see Roach *et al.* (1958). Actually, our present knowledge of nightglow-layer heights is based largely on their direct observation from rockets that traverse the layers.

References

Barbier, D.: 1957, *Compt. Rend.* **244**, 1945.
Barbier, D.: 1958, *Ann. Géophys.* **14**, 334.
Chamberlain, J. W.: 1961, *Physics of the Aurora and Airglow*, Academic Press, New York.
Chapman, S.: 1931, *Proc. Phys. Soc. (London)* **43**, 26; 483.
Elvey, C. T.: 1950, *Am. J. Phys.* **18**, 431.
Hennes, J. and Dunkelman, L.: 1966, *J. Geophys. Res.* **71**, 756–757.
Meinel, A. B., Negaard, B. J., and Chamberlain, J. W.: 1954, *J. Geophys. Res.* **59**, 407.
Rayleigh IV, Lord (R. J. Strutt): 1919, *Astrophys. J.* **50**, 227.
Rayleigh IV, Lord (R. J. Strutt): 1921, *The London Times*, 6c, 27 Jan.
Rayleigh IV, Lord (R. J. Strutt): 1930, *Proc. Roy. Soc. (London)* **A129**, 458.
Rayleigh IV, Lord (R. J. Strutt) and Jones, H. S.: 1935, *Proc. Roy. Soc. (London)* **A151**, 22.
Reed, E. I., Fowler, W. B., and Blamont, J. E.: 1973, *J. Geophys. Res.*, in press.
Rishbeth, H. and Garriott, O. K.: 1969, *Introduction to Ionospheric Physics*, Academic Press, New York.
Roach, F. E., Megill, L. R., Rees, M. H., and Marovich, E.: 1958, *J. Atmospheric Terrest. Phys.* **12**, 171.
Roach, F. E. and Smith, L. L.: 1967, in B. M. McCormac (ed.), *Aurora and Airglow*, Reinhold Publishing Corp., New York.
Tohmatsu, T. and Roach, F. E.: 1962, *J. Geophys. Res.* **67**, 1817.
van Rhijn, P. J.: 1921, *Publ. Astr. Lab. Groningen*, No. 31, pp. 1–83.
Yntema, L.: 1909, *Publ. Astr. Lab. Groningen*, No. 22, 1–55.

0125

PLATE V ROBERT JOHN STRUTT, LORD RAYLEIGH IV

1875–1947

Sons of renowned fathers often have trouble establishing their own legitimate claim to fame, particularly when they are working in the same field, as did the Rayleighs. Fortunately, mutual respect and affection overrode rivalry in this family, and Robert John Strutt built capably on concepts developed by his father, Lord Rayleigh III, branching out into creative research of his own as he matured. His major contributions in the aspects of physics of the atmosphere led to his becoming known as 'the airglow Rayleigh'.

Born on August 25, 1875, Robert John Strutt was the eldest child of Lord Rayleigh III and his wife, the former Evelyn Balfour. Strutt's formal schooling included The Grange, which his Balfour uncles had attended; Eton, in which his interest in and experience with science expanded but where he diligently avoided making friends and participating in sports; and Trinity College, Cambridge, which he entered in 1894. There he started in Mathematics but soon found the Natural Sciences more to his liking. In this latter area at that time, emphasis was on experimental physics, which was taught at Cavendish Laboratory, then directed by J. J. Thomson, a former pupil and close friend of Rayleigh III. There Strutt learned laboratory skills – such as lathe turning and glass blowing, which were necessary because very little ready-made equipment was available – in addition to developing his considerable talents in designing and carrying out experimental problems and devising techniques and equipment to obtain data. During all his school years, he corresponded enthusiastically with his father about the latter's experiments, eagerly inquiring as to what had happened, and how, and why.

Perhaps it was his extensive correspondence – first with his father and later with other scientists and intellectual contemporaries – or perhaps an inborn talent, but in any case, Robert Strutt was a fluent writer and, throughout his life, held in high regard the ability to write concisely and lucidly. His scientific papers were models of excellent exposition, and his frequent forays into more widely read publications – such as letters to the editor of the *London Times* and statements to the House of Lords on matters having scientific aspects – were phrased for clear understanding by the layman who would be their audience. He also wrote two outstanding books – one a biography of his father, Lord Rayleigh III, and the other a biography of J. J. Thomson, his teacher.

Strutt was elected a Fellow of Trinity College in 1900, just a year after the first of his many scientific papers was published in the *Proceedings of the Royal Society*. In 1905, when only 29, he was elected a Fellow of the Royal Society. That was a momentous year for him, as he married Lady Hilda Clements, daughter of the 3rd Earl of Leitrim of Donegal, Ireland. They lived in Cambridge until 1908, when Strutt was appointed Professor of Physics at the Imperial College of Science and Technology in London.

Probably because of his father's work with photography – in the days of exacting requirements of wet collodion plates – Strutt became interested in photography and acquired skills that later were invaluable to many people. During World War I, he learned that the healing of shrapnel-wounded soldiers in a nearby

Lord Rayleigh IV

(Photograph courtesy of John H. Howard, Air Force Cambridge Research Laboratories, Laurence G. Hanscom Field, Bedford, Massachusetts.)

hospital was impeded because the doctors could not locate the shrapnel splinters to remove them. Volunteering his services, he took his equipment to the hospital, X-rayed the patients in their wards, developed the plates in a temporary darkroom nearby, and turned the plates over to the surgeons, who could then remove the remaining splinters. Recognizing a more general need for this service, he volunteered to carry out the same work with the British forces in France and, having moved his family to Terling, the family estate, he was sent to France in June 1915. He was soon recalled to London, however, to serve as a member of a panel working with the Inventions and Research Board to assess the equipment requirements of Admiralty Departments and to evaluate suggestions and inventions submitted to fill those needs. He continued in this capacity until the end of the war.

In April 1919, Lady Hilda, who had been in poor health for several years, died; this loss was a severe blow to Strutt and left him with the care of their four children – three sons and a daughter.

Only a few months after his wife's death, Strutt was again saddened, by the loss of his father, Lord Rayleigh III; at that time, the title, with its privileges and responsibilities, descended on his shoulders. Officially now the master of Terling, he instituted a modernization program, installing an electrical system through the buildings – only his father's laboratory had been served by electricity prior to that time.

Late in 1919 Strutt, now Lord Rayleigh IV, met a war widow, Mrs Kathleen Cuthbert, through mutual friends; they were married in July 1920. Mrs Cuthbert had four children also, of about the same ages as Rayleigh's, and they later had one son, giving them a total of nine children. Rayleigh acquired a car with a special body that accommodated all eleven members of the family, for short trips around the countryside. For more extensive journeys, the family traveled by train and hired bus.

For some time Rayleigh had been interested in the light of the night sky – his father's early paper on scattering of light of the daytime sky may have given him the initial impetus. In 1919 he published his first

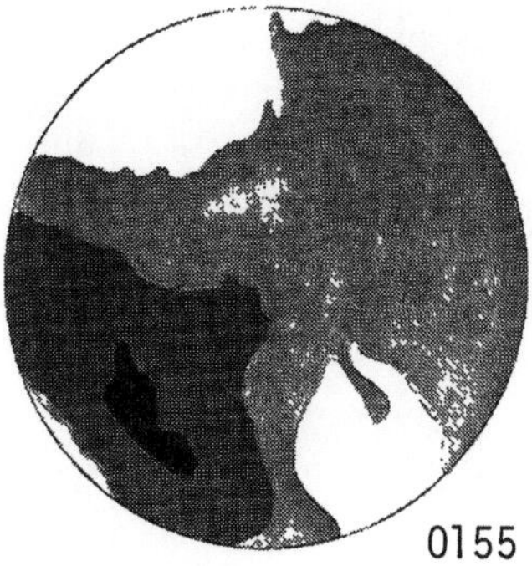

paper on the subject, titled 'Polarization of the Night Sky' (Rayleigh IV, 1919), which was followed over the next 16 years by numerous additional papers on more intensive research, covering observations and measurements of the night-sky spectrum. His early-1920s observations by photographic and visual photometry of the light of the night sky gave quantitative data on time variations of intensity in different spectral regions. On the basis of the 5577 Å green line present even during periods of no auroral activity, Rayleigh was convinced that there was a terrestrial component of the night-sky light that was not auroral in origin, and he designated this component as the nonpolar aurora. In 1930, he presented his paper expressing the radiance of the green (airglow) line in absolute quantal units (Rayleigh IV, 1930), which subsequently resulted in the adoption of the *rayleigh* in his honor (see Appendix 2-B). In the early 1930s he worked with H. Spencer Jones, then Astronomer Royal, further refining his analysis; their joint paper (Rayleigh IV and Spencer Jones, 1935) was his final published contribution in this field.

Rayleigh was widely recognized as a skilled experimenter, ingenious at designing simple yet definitive experiments and devising techniques and equipment to test his theories. His conclusions were, therefore, firmly based and earned for him the respect and recognition of his scientific contemporaries over his half-century career.

Of his 321 published papers, more than half dealt with his extensive original research on gaseous phenomena; his valuable contributions to airglow knowledge comprise a relatively small proportion of that half. His remaining papers covered a wide range of subjects, to many of which he also made significant contributions. More than 50 years ago, Rayleigh published a paper of interest in the light of today's concerns on the part of environmentalists – on oil spills in the oceans and their potentially devastating effects (Rayleigh IV, 1921) – and in 1938 he wrote an article, unpublished, expressing his concern on overpopulation (Rayleigh IV, Ms.).

Thus Rayleigh IV was not a scientific recluse – he was an active participant in the contemporary English scene, enjoying family and social activities and accepting the responsibilities of public service, as well as exerting his scientific talents throughout his life. But he was a scientist to the end – he worked 10 hours in his laboratory on the day before his death, at age 72, in 1947.

CHAPTER 5

DUST-SCATTERED STARLIGHT – THE DIFFUSE GALACTIC LIGHT

Integrated starlight, zodiacal light, and nightglow – which have been discussed to this point – are the major contributors to the terrestrial night sky light. The resolution of the total light into its principal components is much facilitated by the fact that each has a specific region of concentration in the sky, such concentration being apparent even to the unaided eye.

Component	Region of concentration
Integrated starlight	The galactic plane
Zodiacal light	The ecliptic, the Sun
Nightglow	The horizon

The spectroscopically discrete nightglow emissions can be differentiated from the background astronomical light by the introduction of spectroscopic resolution either in the form of photometers with narrow filters, whose peak transmission is centered on the emission feature, or by spectrometers, with resolutions in the 1 to 10 Å range. The nightglow continuum, however, is not separable from the astronomical components by spectroscopic methods; therefore, we must here return to the technique of the preferential region of concentration for its analysis.

The active workers in the photometry of the night sky have developed a feeling for the uncertainties in the published tables, such as Table 3-I in this volume. In particular, the percentage-error bars associated with the faint regions of the zodiacal light (near the ecliptic poles) are probably greater than in the brightest regions.

The purpose of bringing up these uncertainties in the accurate evaluation of the *major* components is to emphasize that serious difficulties are necessarily associated with the estimation of any minor contributor to the night-sky light, such as the so-called diffuse galactic light, a glow associated with the concentration of stars and the scattering of starlight by the dust in the Milky Way.

In an early overall study of the night sky, Elvey and Roach (1937) reported an excess of light in the Milky Way after allowing for that due to the summation of the contribution of the unresolved stars in the photometric observations. A theoretical study by Wang (1936) was published while the observational paper was in press, predicting that interstellar dust co-existing with the stars should scatter the starlight much as the interplanetary dust scatters sunlight to produce the zodiacal light. The obvious independence as well as the quantitative agreement of the prediction and the observations led to general acceptance of the existence of the diffuse galactic light.

In retrospect, it is evident that the numerical agreement of the two studies may have been somewhat fortuitous, since the uncertainties in the observational analysis were significant. However, the existence of the excess galactic light has been subsequently confirmed by several investigators.

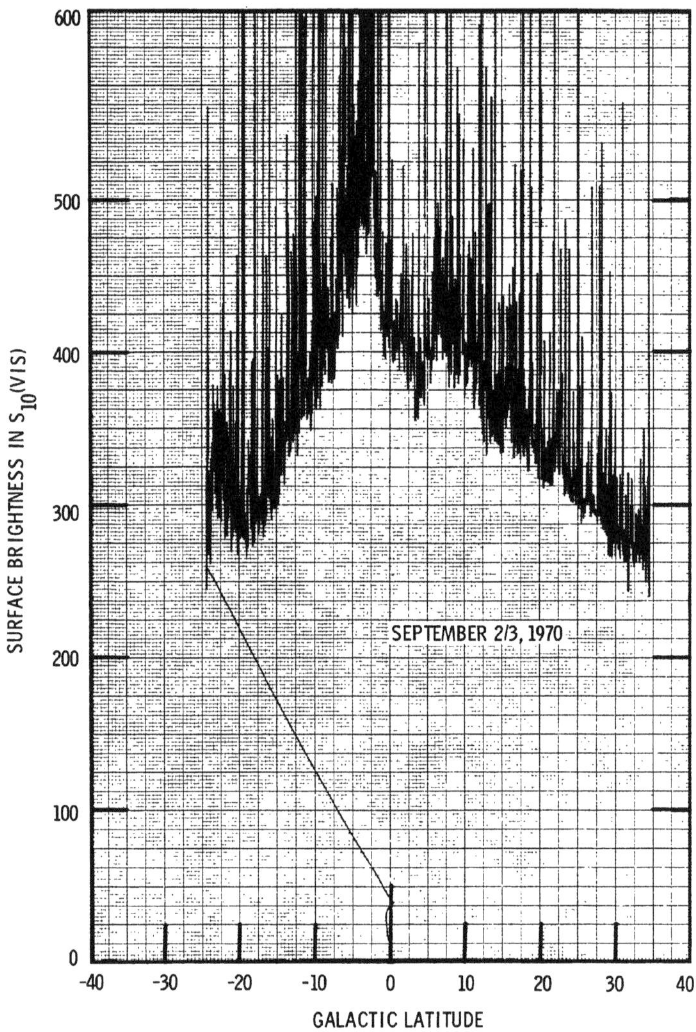

Fig. 5-1. Plot of 2800 successive readings of sky brightness across the Milky Way with a fixed telescope. The Earth's rotation sweeps the astronomical sky across the field of view and, as individual stars enter it, discrete spikes appear on the plot. The underdrawn envelope tracing the lower points of the star deflections represents essentially the upper level of the diffuse galactic light superimposed on a general background of zodiacal light and nightglow.

The diffuse galactic light is brightest, in absolute terms, in the Milky Way, where both the interstellar dust that scatters the light and the concentration of stars themselves are at a maximum. If a large field of view is used, the contribution to the photometric reading from the integrated starlight is very strong, as is evident from even a casual examination of the heavy concentration of stars (Figure 2-2b) in the star-rich regions of the Milky Way. Many of the historical papers dealing with the problem have attempted to allow for the integrated light from the stars by a direct or, more often, indirect application of star-count data such as discussed in Chapter 2. The uncertainty introduced into the analysis by an application of this differential-residual method is severe, and we shall concentrate in this chapter on some recent results in which the employment of astronomical telescopes has permitted the use of small fields for which it is possible to make readings between most of the stars and thus minimize the correction for the general starlight.

To illustrate the nature of the problem, we refer to Figures 5-1 and 5-2. Figure 5-1 is a plot of a series of some 2800 digitized readings made with a photometer attached to a 16 in. telescope at Kitt Peak, Arizona. The telescope was fixed in position at a

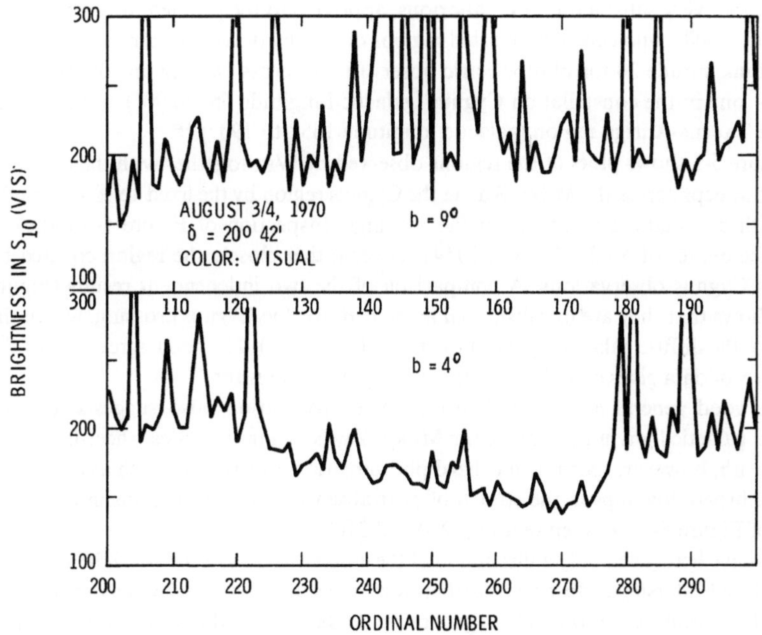

Fig. 5-2. Expanded plot of observed sky brightness of a small portion of the domain shown in Figure 5-1 – between galactic latitudes (*b*) 9° and 4° – using a fixed telescope. The readings were taken at 15 s intervals through an entire night; 200 contiguous readings are shown in this plot. Star deflections again appear as discrete spikes. In this case, the zodiacal light and nightglow have been subtracted out and the curve represents primarily the diffuse galactic light, with a small but not trivial contribution from interstellar light from stars too faint to record individually.

declination of 20°42′ and an hour angle of 2 hr W. The observations have been ordered in galactic latitude, and the brightness as the Milky Way traverses the field is apparent. The evaluation of how much of the maximum is due to diffuse galactic light has been simplified by the use of a relatively small field*. Figure 5-2 gives the detailed plot of 200 contiguous readings of a small part of the region covered in Figure 5-1, showing the detailed photometric structure of a portion of the Milky Way. In Figure 5-2 the zodiacal light and the airglow have been subtracted out, in contrast to Figure 5-1, for which much of the general background light is due to these contributors.

In a traverse of the Milky Way the concentration of stars maximizes, as does the background light, as shown in Figure 5-1, in which the numerous star deflections stand out above the underdrawn envelope. The star-deflection density was used to evaluate the correction necessary to allow for the residual light from very faint stars. A striking feature of the analysis is that the region of the slight dip in the readings near galactic latitude +4° is actually brighter than its environs because the star correction here is almost trivial. The method outlined is now being used with fields so small that the correction for residual integrated starlight is trivial.

Another very satisfactory, but laborious, approach to the problem was introduced by Witt (1968), who carefully selected regions of the sky devoid of stars brighter than visual magnitude 20 for photoelectric observations across two regions of the Milky Way – one in the constellation Cygnus (galactic longitude 65° to 90°) and the other in the Taurus-Auriga region (galactic longitude 165° to 190°).**

Figure 5-3 shows plots of the regions observed by Witt together with the traverses of two sweeps across the Milky Way in the Cygnus region by the fixed-telescope method with automatic digitization of the readings. Inspection of Figure 5-3 indicates that the curves of $\delta = 31°58'$ and 20°42′ cover rather closely the region covered by Witt's Cygnus observations. A comparison of the two independent results (Figure 5-4) shows that they are in substantial agreement for the Cygnus crossing, where the peak of the diffuse galactic light is more than 100 S_{10}(vis). The good agreement of the two sets of data gives confidence in the validity of both results.

The good general agreement between these observations also suggests that the diffuse galactic light in this part of the Milky Way is of a more-or-less sheet-like character with, however, sporadic localized photometric perturbations such as evidenced by the broad, low pips in the stretch of near-absence of star deflections in the lower part of Figure 5-2 (between readings 230 and 270).

We have here given only a discussion of the empirical facts of the starlight scattered by diffused interstellar dust and, furthermore, the sample is an extremely small fraction of the total. It is impossible to generalize on the basis of this amount of evidence, but in the next chapter we propose to discuss some of the properties of the star-

* The reader should refer to the original paper for a description of the technique used (Roach *et al.*, 1972).
** Note that Figure 2-6 indicates that, in the case of Cygnus, we are looking along a spiral arm and that, in the case of Taurus-Auriga, across two arms.

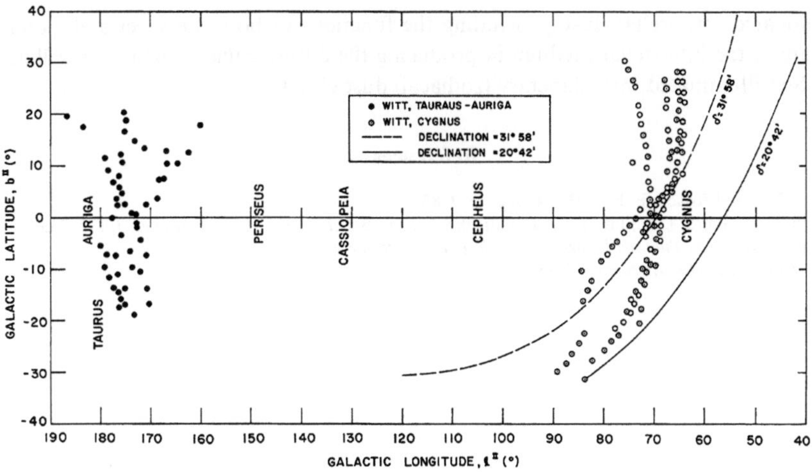

Fig. 5-3. Plot of portions of the sky that have been studied for the presence of diffuse galactic light. The dot and dot-circle symbols represent regions examined by Witt (1968). Two regions covered by the fixed-telescope technique (Roach *et al.*, 1972) are indicated by the continuous plots.

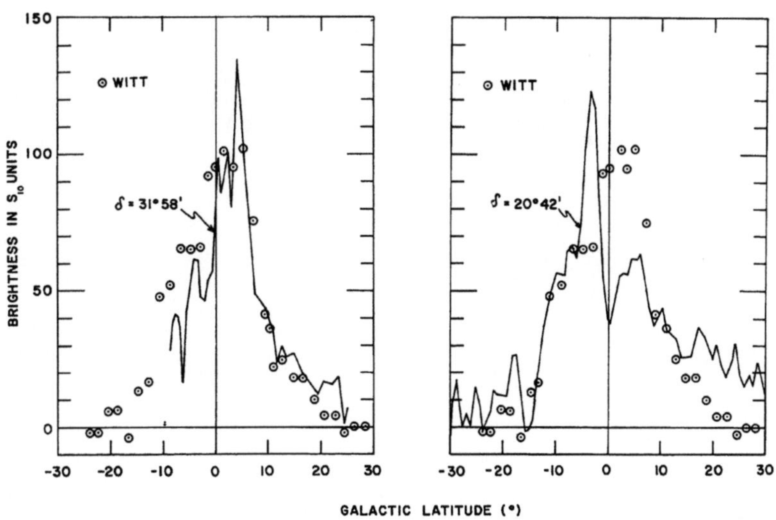

Fig. 5-4. Comparison of independent measurements of diffuse galactic light by Witt (dot-circles) in the Cygnus region of the sky, and by Roach *et al.* (1972) (solid lines) by the fixed-telescope technique. In the latter, the effect of direct starlight is allowed for in the reduction of the data.

illuminated interstellar dust – including the functions of both the sheet and cloud nature of the interstellar medium in producing the diffuse galactic light – as well as the Sun-illuminated interplanetary (zodiacal) dust cloud.

References

Elvey, C. T. and Roach, F. E.: 1937, *Astrophys. J.* **85**, 213.
Roach, F. E., Smith, L. L., Pfleiderer, J., Batishko, C., and Batishko, K.: 1972, *Astrophys. J.* **173**, 343.
Wang, S. K.: 1936, *Publ. de l'Observatoire de Lyons, fasc. 19* **1**.
Witt, Adolf N.: 1968, *Astrophys. J.* **152**, 59.

CHAPTER 6

DUST – INTERPLANETARY AND INTERSTELLAR

In the physical universe, perturbative clumpings of material include the entire spectrum of sizes from the subatomic to the metagalaxy. The span of masses from the electron to our Galaxy goes from 10^{-27} to 10^{44} gm, or 71 powers of 10. Between these two extremes, Nature displays a comprehensive array to the astronomer, as listed in Table 6-I.

TABLE 6-I

Array of particulate matter

Species	Mass range (gm) (logarithmic)
Electrons	−27
Atoms	−24
Molecules	−23
Dust grains	−12 to 0
Pebbles, rocks, boulders (meteors, meteorites)	0 to 15
Comets	16 to 19
Asteroids	19 to 22
Moons	22 to 26
Planets	26 to 31
Stars	31 to 35
Galaxies (10^{11} stars like the Sun)	44

The fact that we are interested in dust in this book is not because it is, in itself, more significant astronomically than the other species, but because of its observability as a contributor to the light of the night sky. The observability of dust is somewhat fortuitous, depending first on its astronomical concentration, and second on the fact that the large number of small dust particles presents a larger cross section to light than do the smaller concentrations of large particles between dust and stars. Thus meteors, comets, asteroids, moons, and planets can be directly observed in the restricted region of the solar system, but their existence in interstellar space or associated with other stars must be inferred by indirect methods. The presence of interplanetary dust is indicated by the zodiacal light (Chapter 3) and of interstellar dust by the diffuse galactic light (Chapter 5). We make a digression to illustrate the statements of this paragraph.

According to studies of the dynamics of our Galaxy (Schmidt, 1965), the rotating system requires a mass distribution from the galactic center out to and beyond the Sun, according to Figure 6-1. Of particular interest is the requirement that at the

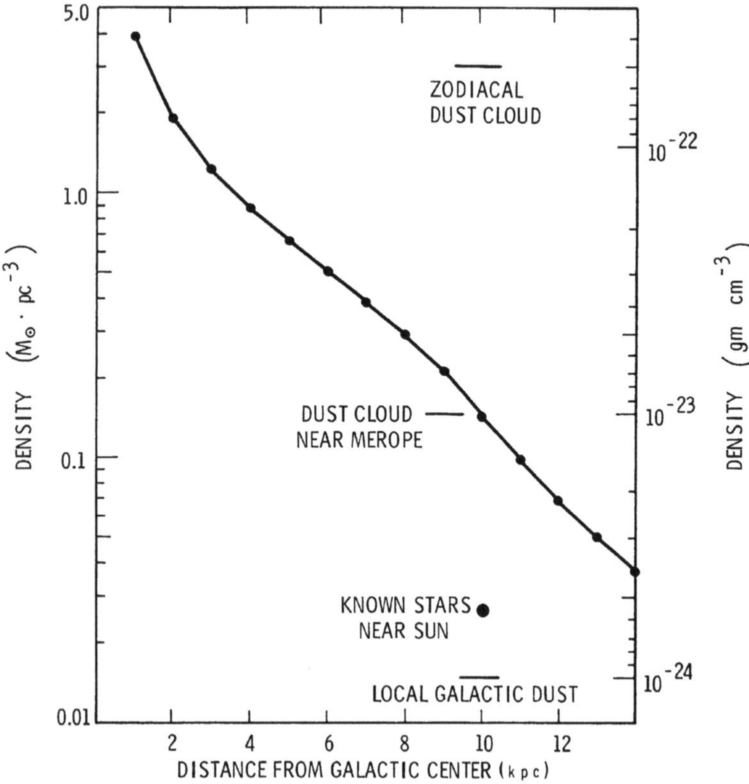

Fig. 6-1. Distribution of mass density in the Galaxy deduced from the dynamics of its rotation according to Schmidt (1965). At the distance of the Sun from the center of the Galaxy (10 kpc) the density is 0.14 $M_{\odot}$ pc^{-3}.

Sun's distance from the galactic center (10 kpc) the density of material be 10^{-23} gm cm^{-3} or 0.14 $M_{\odot}$ pc^{-3}*.

Now, counts of stars (see Chapter 2) referred to the volume density in the astronomical vicinity of the Sun account for 0.02 $M_{\odot}$ pc^{-3}, or only 14% of that required by the galactic dynamicists. So we have a detective problem of astronomical proportions to locate the missing mass. It may be speculated that the several components listed in Table 6-I are distributed throughout interstellar space. A model based on such a speculation is presented in Figures 6-2 and 6-3. Figure 6-2 is basically an *interpolation* between the known space density of observable stars on the high-mass side of the plot to a model of the galactic-dust space density on the low-mass side. The implication of the plot is that *all* of the objects of Table 6-I occur throughout our part of the Galaxy. In Figure 6-3 we show the mass-density contributions (on a linear scale) according to

* $M_{\odot}$ = solar mass.

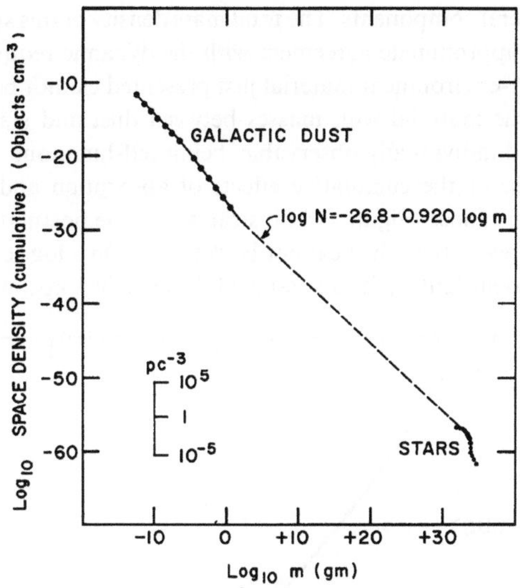

Fig. 6-2. An interpolation between the known space density of observable stars in the vicinity of the Sun and the interstellar dust cloud. The interpolated region includes all objects between these two extremes.

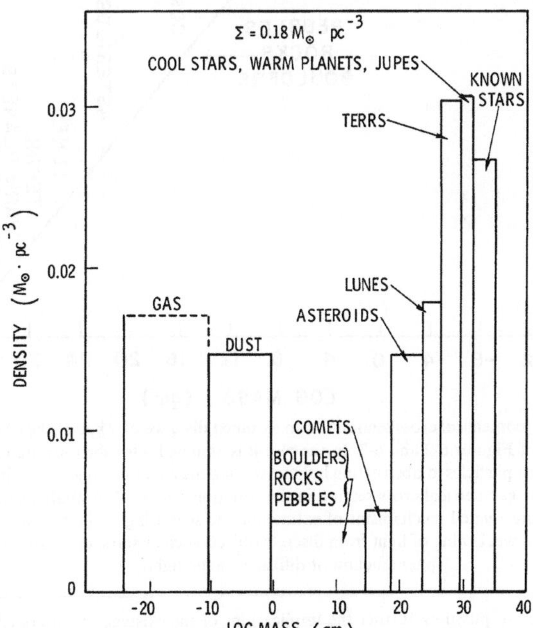

Fig. 6-3. A speculative model of the distribution of material in the vicinity of the Sun, based on the interpolation of Figure 6-2.

the masses of the several components. The total mass density of this speculative model is 0.18 $M_\odot$ pc^{-3}, in approximate agreement with the dynamic requirement*.

The model of solar-environment material just presented cannot be tested observationally insofar as the material with masses between dust and stars is concerned. Stars are directly and individually observable, being self-luminous. Interstellar dust is observable because of the cumulative effects of absorption and scattering over great astronomical distances. Figure 6-4 illustrates why the postulated astronomical debris with masses greater than dust cannot be detected. On a log scale is plotted the effective extinction coefficient, τ, for an assumed distance of 1 kpc, where τ is defined

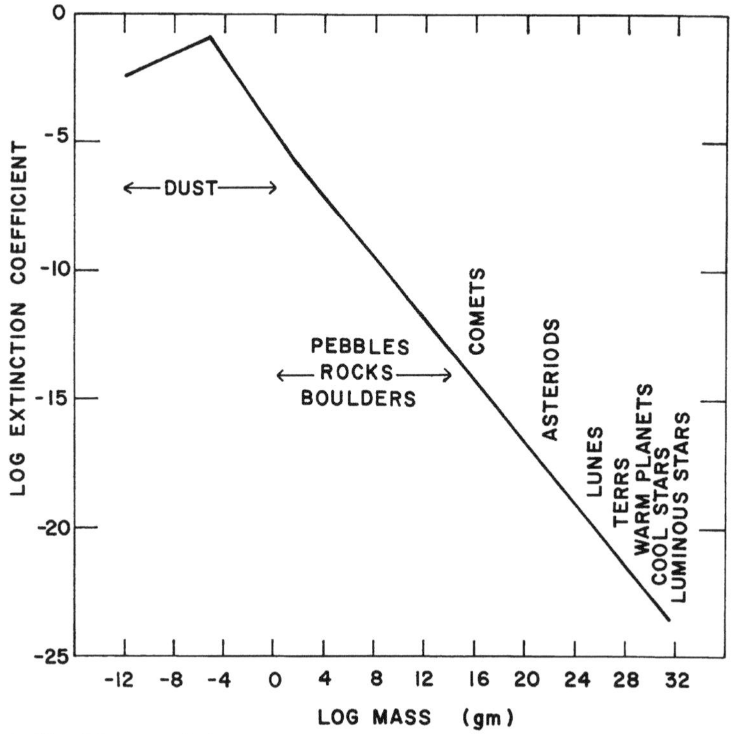

Fig. 6-4. Plot of the geometrical cross section for an assumed distance of 1 kpc of the objects represented by the interpolations of Figures 6-2 and 6-3. In the plot it is assumed – for the purpose of illustrating the superiority of small dust particles in obstructing light traversing the medium – that the extinction coefficient τ is proportional to the geometrical cross section in the equation $I/I_0 = e^{-\tau}$. Actually, small dust particles are observable from interrelated mechanisms of extinction and scattering of light traversing the medium, the extinction causing a weakening of light from discrete objects such as stars, the scattering producing the phenomenon of diffuse galactic light.

* The reader interested in pursuing further the implications of the existence of interstellar debris might refer to Oort (1950), who considered the matter of a swarm of comets in the solar domain (halfway to Alpha Centauri), and Shapley (1962), who speculated on the general existence of cool stars and crusted planets in interstellar space.

by the following equation:

$$I/I_0 = e^{-\tau} \approx e^{-\pi s^2 nl} = e^{-\sigma nl}. \tag{6.1}$$

The symbols are: I = the intensity of the transmitted light through the region, I_0 = the intensity of the original beam, s = the radius of the particle (cm), n = the space density of the particles (cm^{-3}), and l = the path distance (1 kpc $= 3.09 \times 10^{21}$ cm). We shall return to this later but wish here to emphasize (1) that as τ approaches unity ($\log \tau \to 0$), as it does for the dust side of the plot, the extinction is significant*, and (2) that, where $\tau \ll 1$, as is the case for the larger-than-dust components, this family of objects cannot be observed by their cumulative extinction effects. Since the latter are not self-luminous, we have an observation gap between dust and stars.

Some amusing calculations of these elusive unobservables may be made. In the solar domain (defined as a sphere with radius halfway to Alpha Centauri), there are predicted six objects as massive as the planet Jupiter and 89 as the planet Neptune. Perhaps we should revise our concept of the morphology of our solar system. Within a sphere *including* Alpha Centauri, the model predicts five stars similar to the famous 'Barnard's star'. Searches for such intrinsically faint stars are, of course, part of the continuing efforts of astronomers.

The purpose of this general discussion of the total budget of astronomical material in our part of the Galaxy is to put into perspective the balance of the chapter, in which we shall discuss the zodiacal cloud of interplanetary dust and then the interstellar dust clouds associated with the diffuse galactic light.

6.1. The Zodiacal Dust Cloud

Referring back to Figure 6-4, we there used the geometrical cross section of the particles as an indication of their power to intercept light. The quantity τ, as so determined and used in Equation (6.1), is certainly valid for the large objects in estimating the extinction or absorption of a light beam traversing the region containing the particles. But the smaller particles cannot be considered as simple opaque objects with respect to the traversing light beam. The critical parameters are: (1) the size of the particle relative to the wavelength of the light, and (2) the index of refraction of the particles with respect to the light. A two-dimensional schematic array of the complex types of analysis is shown in Figure 6-5, taken from van de Hulst (1957). We have already encountered in Chapter 1 the case of Rayleigh scattering applicable in cases where the scattering particles of the atmosphere (molecules) are very small compared with the wavelength of the light being scattered. In general, the problems associated with interactions between the zodiacal dust cloud and sunlight and between the interstellar dust cloud and starlight are concerned with particles *not* small with respect to the wavelength of the light, a domain usually referred to as 'Mie scattering' after the investigator who concentrated on the problem.

* For example, $I/I_0 = 0.725$ for $\tau = 0.32$.

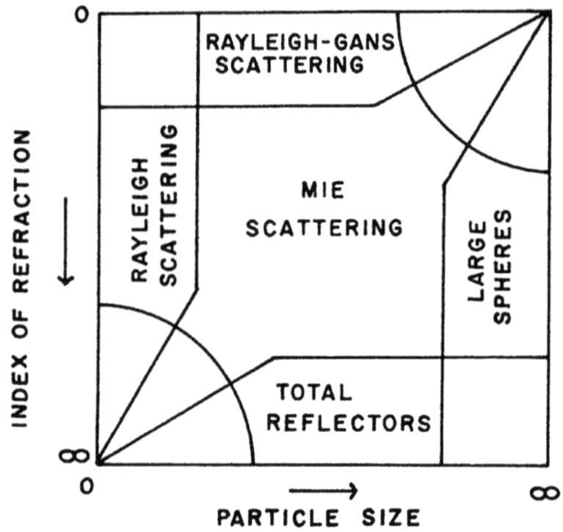

Fig. 6-5. Schematic diagram illustrating the complex of interactions between light and particles. The size of particle increases along the abscissa from very small to very large. The ordinate represents the index of refraction of individual particles, increasing downward (Adapted from Figure 20, van de Hulst, 1957: 132).

An understanding of the nature of the zodiacal cloud is approached not only by an interpretation of the photometric data on the zodiacal light but through such apparently diverse inquiries as: meteor studies (both optical and radar), meteorite investigations including their impact craters, and dust collecting and counting from space vehicles. All of these disciplines lead toward a knowledge of what has been called the 'meteoritic complex', small particles in interplanetary space, ranging in size from dust grains to small asteroids. According to Millman (1967), the basic character of the complex is given by a knowledge of its space density, velocity, and size distribution. Each of these properties is a function of distance from the Sun, time, and clumping. Finally, it is desired to know the nature of the material, such as composition, density, structure, hardness, and strength. It is obvious that such a large number of parameters cannot be determined explicitly from a restricted observational domain, or even from the ensemble of all the observational approaches. A specialist in any one of the disciplines makes a partial contribution to the interpretation, but a generalist is needed for the complete diagnosis.

The observations of the zodiacal light (and the gegenschein) help in estimating the space density near the Earth and, as a function of distance from the Sun, the size distribution of the particles, and their possible clumping.

There is a wide diversity of 'models' of the zodiacal cloud, as evidenced by the entries in columns 3 and 4 of Table 6-II from Bandermann (1967), which represent the extremes that have been postulated to account for the observational facts of the zodiacal light. A generally accepted definitive model between these extremes is not available, and we have chosen to illustrate the probable *general* nature of the interplanetary dust by the use of the parameters shown in column 5. Figure 6-6 shows the

TABLE 6-II

Comparison of models of zodiacal cloud

Item	Parameter	Bandermann A	Bandermann B	Illustration[a] (this book)
1	Typical size or radius, s	'Large'; several $\times 10^{-3}$ cm	'Small'; $<10^{-4}$ cm	10^{-2} cm
2	Particle space density at 1 AU in ecliptic plane	10^{-21} gm cm^{-3}	10^{-23} gm cm^{-3}	2×10^{-22} gm cm^{-3}
3	Variation of space density with distance from Sun, r	$N \propto r^{-1.5}$	$N \propto r^{+0.5}$	$N \propto r^{-0.5}$
4	Distribution index, p; $N \propto s^{-p}$	'Steep'; $p>4$	'Flat'; $p<3$	$p=4$ for $s>10^{-2}$ cm $p=1.35$ for $s<10^{-2}$ cm
5	Composition	Metallic	Stony (dielectric)	
6	Average particle density, ρ, gm cm^{-3}	'Compact'; >3 for stones; 7.8 for metals	'Fluffy'; <0.1	1.0 for $s<10^{-2}$ cm 0.44 for $s>10^{-2}$ cm
7	Albedo, a	>0.1	<0.05 ('sooty')	
8	Shape	Nearly spherical	Needles; platelets	
9	Equilibrium electric potential	Small, + (a few volts)	Large, + or −	
10	Typical eccentricity of dust orbits, e	$\leqslant$ asteroids	Large, $e>0.5$	
11	Average inclination of dust orbits	A few degrees	$>20°$	
12	Typical geocentric speed	$5(^{+10}_{-2})$ km s^{-1}	$\gg 15$ km s^{-1} or $\ll 5$ km s^{-1}	

[a] The examples chosen for illustration here are from Whipple (1967: 420, Table 3) for items No. 1, 2, 4, and 6, and from Gillett (1966) for item No. 3.

percentage distribution by mass according to the model along with a plot on a logarithmic scale of the 'age' of the particle in each mass regime according to Whipple (1967). Figure 6-7 shows the space density of the particles vs their radius according to the same model. In Figure 6-8 the relative geometrical cross section of the ensemble of particles is shown plotted against the mass on a logarithmic scale.

Finally, in Figure 6-9 is displayed the particulate space density in the ecliptic plane vs distance from the Sun on the assumptions: (1) a space density of 2×10^{-22} gm cm^{-3} at the Earth's distance of 1 AU, and (2) variation according to $r^{-0.5}$, where r is the distance from the Sun.

Immediately obvious is the fact that the present cloud cannot be a residual of the primordial material of the solar system. The mean life of the present cloud – 80000 yr (Whipple, 1967; also see Figure 6-6) – is a trivial fraction of the age of Earth, and if it represents a steady state there must have been many thousands of turnarounds of zodiacal material during the history of our planet. Another generalization that can be made is that it is *not* possible that the zodiacal cloud is simply a part of the interstellar dust cloud. The space density of the zodiacal cloud is about 20 times the *total* required in the solar environment for dynamical stability of the rotating Galaxy (see Figure 6-1).

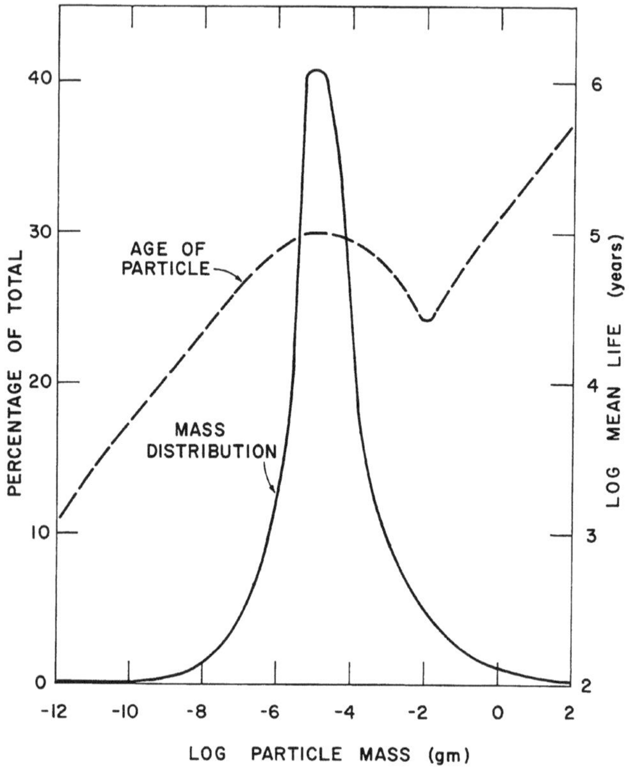

Fig. 6-6. The lifetime of the zodiacal cloud (according to Whipple, 1967).

If the present zodiacal cloud may be considered quasi-stable over a few centuries, it is pertinent to ask how it is maintained against the steady losses due to light pressure, spiraling-in (Poynting-Robertson effect), sublimation of particles, losses by solar wind, grinding by collisions, etc. We look for mechanisms that can supply some 10 to 20 metric tons of material per second. Two potential sources are the asteroids and the comets. The loss rate which has to be replenished corresponds to the grinding up of one asteroid such as Ceres (mass $\approx 7 \times 10^{23}$ gm) in about 10^{10} yr. Whipple has calculated that the mass loss of two periodic comets is comparable with the requirement for maintaining the zodiacal cloud – Halley's comet, 5 tons s^{-1}; Encke's comet, 3.5 tons s^{-1}. It is possible that *both* asteroidal and cometary contributions to the cloud are significant.

Of particular importance in evaluating the physical nature of the dust particles is the measurement of the polarization of the zodiacal light (see Figure 3-3). Early interpretations attributed this polarization to a concentration of electrons permeating the zodiacal cloud. That scattering from particulate material in the dust-grain category could produce polarization has led investigators into vigorous experimental and

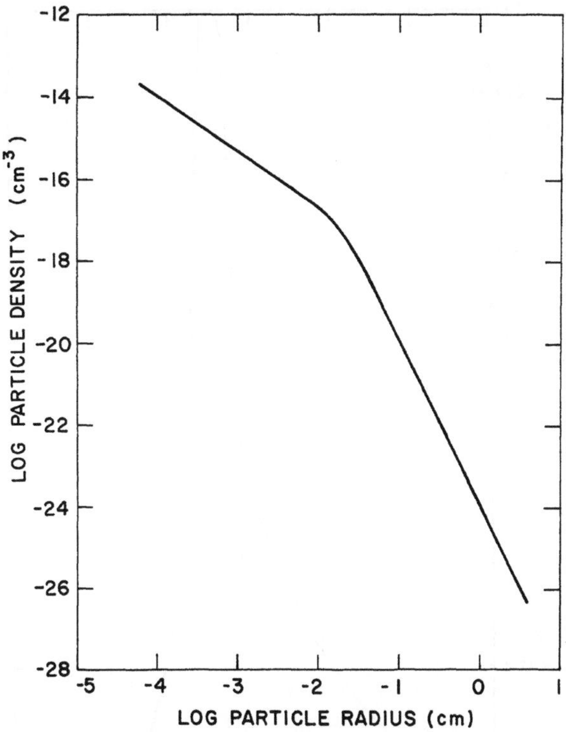

Fig. 6-7. Space density of the zodiacal dust cloud with respect to the particle radius (according to Whipple, 1967).

computational attempts to rationalize the observations with predominantly dusty models.

Temporal and/or spatial clumping of the interplanetary material has been actively considered, especially since 1961, when Polish astronomer Kordylewski (1961) reported observations of photometric 'perturbations' near the so-called LaGrangian points of equilibrium in the Earth-Moon gravitational system. The mathematician LaGrange, in studying a restricted three-body problem, called attention to the fact that there are, in the plane of the orbit of two bodies, five regions of quasi-stability (Figure 6-10). Of particular interest are the points L4 (preceding) and L5 (following) at the apices of equilateral triangles formed by the two bodies. If there are no perturbing forces such as the presence of another massive body, then a nonmassive body in the L4 or L5 position will tend to be gravitationally caught in the equilibrium positions, perhaps performing oscillations (librations) around them. Applying this idea to the Sun-Jupiter system, astronomers have confirmed that there are more than a dozen asteroids (usually called Trojans and Greeks) at distances from the Sun equal to that of Jupiter and relatively close to the equilibrium positions.

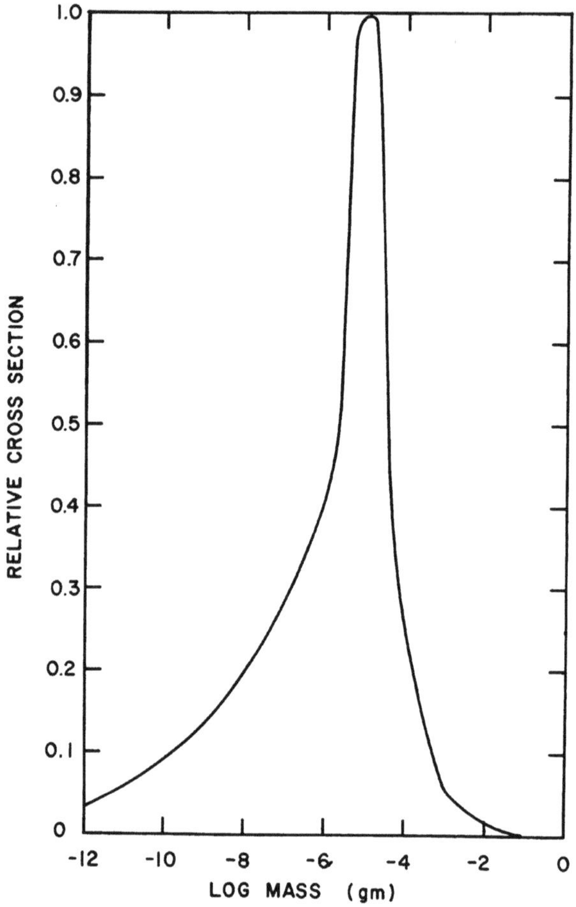

Fig. 6-8. The zodiacal cloud according to Whipple (1967), showing the relative geometrical cross section vs the logarithm of the mass.

A careful search of one of the equilibrium positions in the Earth-Moon system for discrete asteroidal objects was unsuccessful, but interest in the possibility of a concentration of dust in these positions resulted from the report of Kordylewsky, and more recently by Simpson (1967). Recent studies suggest that the photometric perturbation resulting from concentration of dust in the Earth–Moon equilibrium positions may be as small as 10% to 20% of that of the gegenschein (Roach, 1973), which could account for the fact that some investigators have not been able to distinguish the effect (see, for example, Roosen, 1968).

The gegenschein has been ascribed to: (1) a phase effect – that is, a preferential *back*scatter of sunlight or zodiacal particles in the region beyond the Earth's orbit, (2) a concentration of particles at the L3 position of the Earth-Sun system, or (3) a

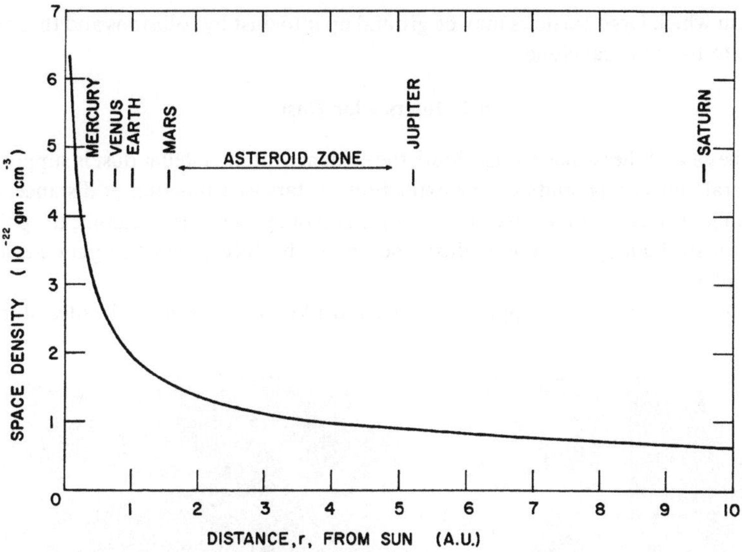

Fig. 6-9. Plot of the space density of the zodiacal cloud on the assumption that it varies according to the −0.5 power of its distance from the Sun.

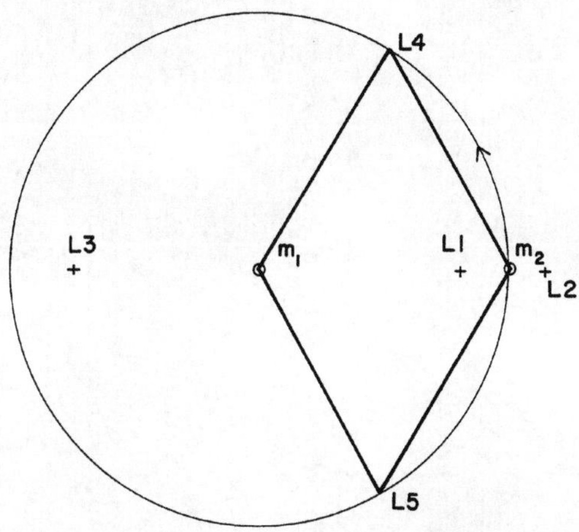

Fig. 6-10. Schematic representation of the quasi-stable regions (LaGrangian points) for a small object in the plane of two large rotating bodies.

combination of these effects. At the time of writing (September, 1973), preliminary reports from the satellite Pioneer 10 indicate that the photometric gegenschein persists well beyond the Sun-Earth L3 point. When the data are completely analyzed, it should be possible to determine whether there is any degree of particulate concentration in the L3 region, as well as in the asteroidal belt. The latter is a possible primary region in which large particles may be ground up into dust by collisions and thus contribute to the zodiacal cloud.

6.2. Interstellar Dust

As in the case of the zodiacal dust cloud, the evidence for interstellar dust is supported by several converging studies – the extinction of stars as a function of distance, the reddening of starlight that traverses either general or specific dusty regions, the polarization of starlight by intervening dust, and finally the diffuse galactic light described in Chapter 5.

The general gross visual appearance of the Milky Way as well as detailed studies

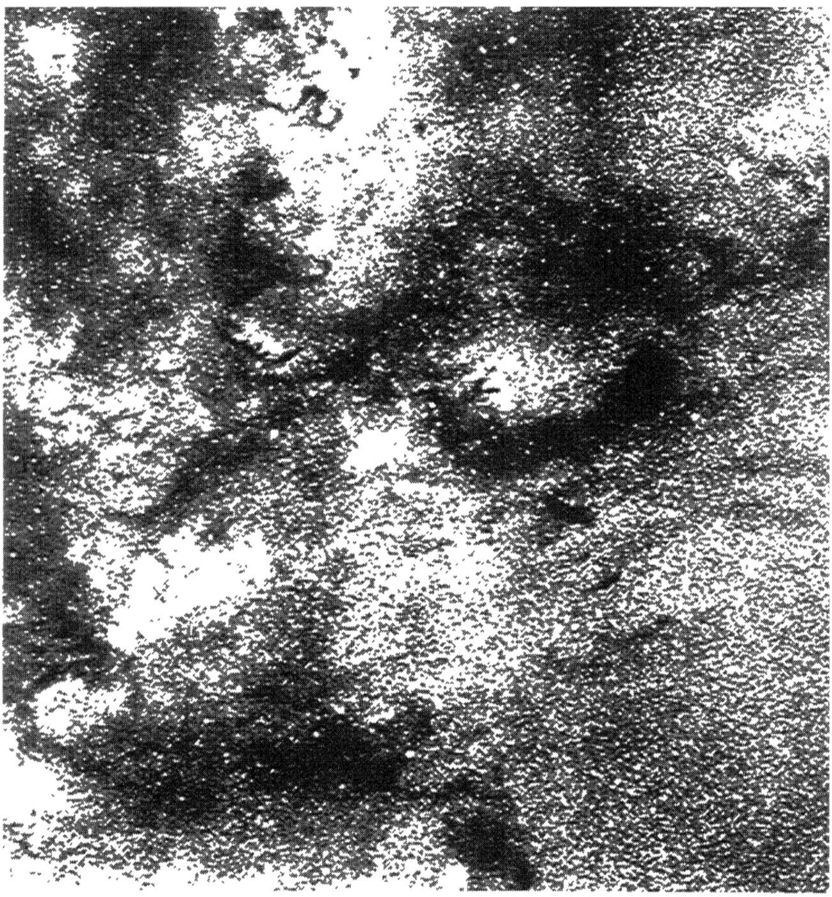

Fig. 6-11. Photograph of a portion of the Milky Way near the star Theta Ophiuchi, showing many 'dark' regions due to intervening dust clouds that obscure the more distant stars.

show the effect of obscuration. The great rift in the Cygnus region is one of the features that catches the eye of the night watcher. Figure 6-11 shows a photograph of a region in the Milky Way in which the irregularities are obviously not due to perturbations in the distribution of stars but rather to the presence of intervening dust, which often has sharp delineations. Some of the very small blobs have been interpreted as regions where stars are in the birth process.

The general transmission of starlight looking outward from the solar system is $I/I_0 = Al = e^{-\tau}$, where I/I_0 is the fractional transmission of a star's light and l is the distance to the star in kiloparsecs, and A is in stellar magnitudes. In the vicinity of the Sun, $A \approx 0.35$ stellar magnitude/kpc. Thus in 1 kpc, starlight is weakened by 0.725. This very considerable diminution of a star's apparent brightness is accomplished by an extraordinarily thin density of dust extending over the vast distance of a kiloparsec (3.09×10^{21} cm).

A somewhat oversimplified calculation will illustrate the space density of dust particles required to give an extinction of 0.35 mag. for a star at a distance of 1 kpc. It is convenient to take advantage of the sharp maximum in the relative cross section of particles in the Whipple zodiacal cloud at a mass of 10^{-5} gm (Figure 6-4) and apply it to the interstellar cloud. Order-of-magnitude calculations can be made using this as a 'standard particle'. With a particle radius of 0.0174 cm, the geometrical cross section of each particle is 9.5×10^{-4} cm^2. A 0.35-mag. weakening (transmission $= 0.725$) calls for

$$I/I_0 = 0.725 = e^{-\sigma n l},$$

where σ = the cross section of each particle, n = the space density of the particle, and l is the length of the path. For $I/I_0 = 0.725$, $\sigma n l = 0.322$, whence

$$\begin{aligned} n &= 0.322/3.09 \times 10^{21} \times 9.5 \times 10^{-4} \\ &= 1.1 \times 10^{-19} \text{ particle cm}^{-3} \; (1.1 \times 10^{-24} \text{ gm cm}^{-3} \text{ for } m = 10^{-5} \text{ gm}). \end{aligned}$$

This corresponds to a space density of $(1.1 \times 10^{-24})/(2 \times 10^{-22}) = 0.005$ times that of the zodiacal cloud. The dust cloud around the star Merope in the Pleiades (Figure 6-12) has been estimated to have a mean density of 10^{-23} gm cm^{-3}, or 10 times that in the general interstellar medium (Figure 6-1). It is of interest to calculate the effects on light transmission for three cases, as shown in Table 6-III for several assumed distances. It is noted that a light beam 1 cm^2 in cross section is greeted by a standard particle in the general stellar domain every 10^{19} cm or 3.3×10^8 s (approximately 10 yr) (Figure 6-13). If such a cloud is in the immediate vicinity of our solar system, stars closer than 10 l-yr (c. 3 pc) could get a 1-cm^2 light signal to us without a single interruption by a standard dust particle until the zodiacal cloud was reached. Of such gossamar stuff is our Galaxy composed.

If the interstellar dust were as dense as the local zodiacal cloud (Figure 6-1), serious diminution of the brightness of even the nearby stars would occur. A star at 1 kpc distance would be completely invisible ($I/I_0 = e^{-60} = 10^{-26}$). A star at even the relatively close distance of 100 pc would be very seriously diminished in brightness ($I/I_0 = 0.0025$).

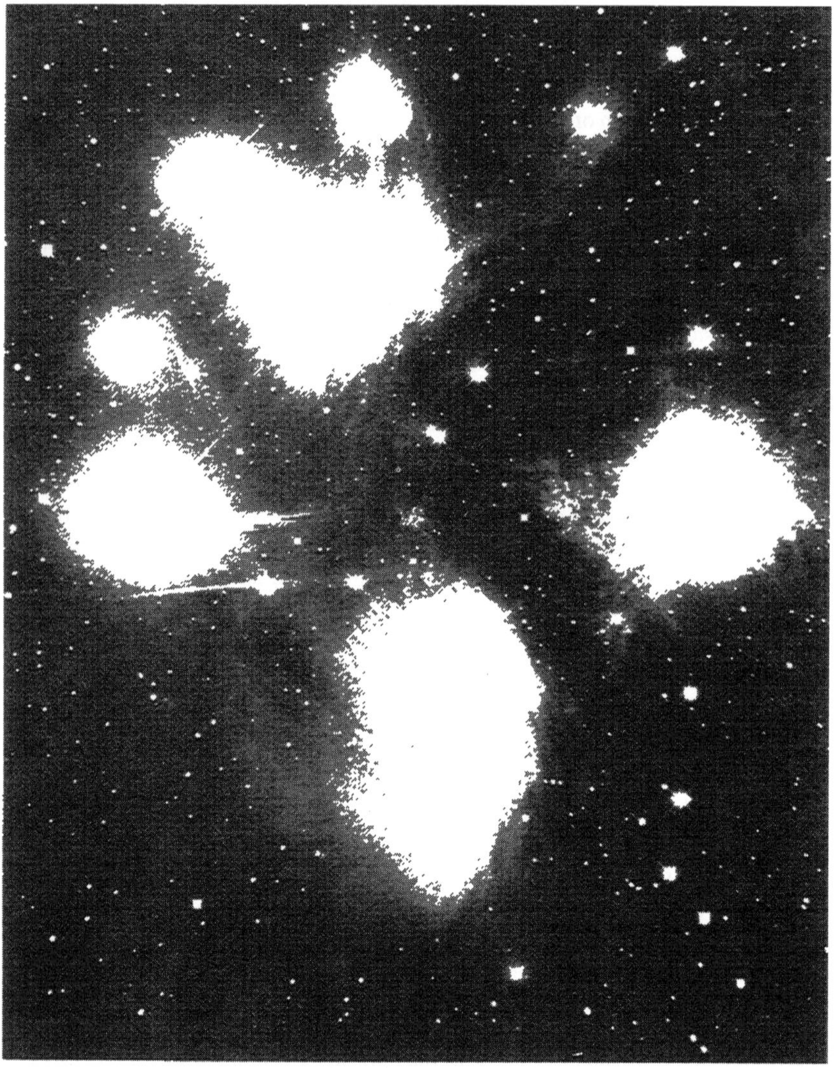

Fig. 6-12. Nebulosity (gas plus dust) around the stars of the Pleiades cluster. The star Merope is embedded in the nebulosity in the lower part of the photograph.

On the other hand, the zodiacal cloud is of very low density in terms of solar-system distances. If its density in the vicinity of the Earth were to persist for 40 AU (to the orbit of Pluto) the weakening of the brightness of a star whose light traversed such a cloud would be only about 1% ($I/I_0 = 0.988$).

The general nature of interstellar-dust parameters based on the above considera-

TABLE 6-III

Extinction due to dust (standard particle, $\sigma = 10^{-3}$ cm^2) at different distances, l ($I/I_o = e^{-\sigma n l}$)

Cloud and its space density	1 AU ($l = 1.5 \times 10^{13}$ cm)			1 pc ($l = 3 \times 10^{18}$ cm)			1 kpc ($l = 3 \times 10^{21}$ cm)			10 kpc ($l = 3 \times 10^{22}$ cm)		
	nl	σnl	$e^{-\sigma nl}$	nl	σnl	$e^{-\sigma nl}$	nl	σnl	$e^{-\sigma nl}$	nl	σnl	$e^{-\sigma nl}$
Zodiacal cloud ($n = 2 \times 10^{-17}$ cm^{-3})	3(−4) 0.0003	3(−7)	0.9999997	60	0.06	0.942	60000	60	0	600000	600	0
Merope cloud ($n = 10^{-18}$ cm^{-3})	1.5(−5) 0.000015	1.5(−8)	1	3	0.003	0.997	3000	3	0.050	30000	30	9(−14)
Interstellar cloud ($n = 10^{-19}$ cm^{-3})	1.5(−6) 0.0000015	1.5(−9)	1	0.3	0.0003	0.9997	300	0.3	0.741	3000	3	0.050

[a] σ = extinction cross section of particle, cm^2.
n = space density of particles, cm^{-3}.
l = path length, cm.

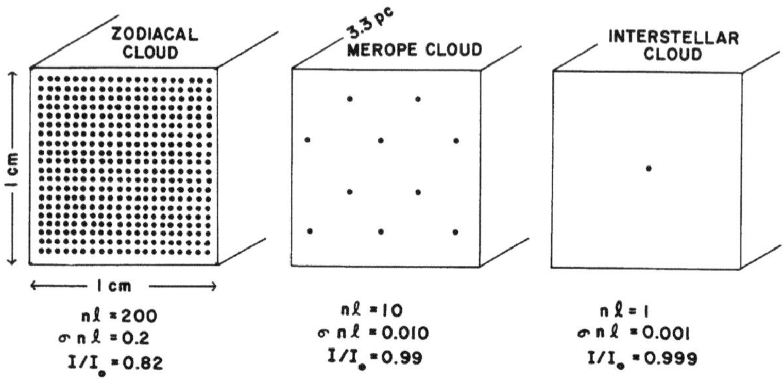

Fig. 6-13. Schematic representation of the relative number of dust particles included within a square-centimeter column extending over a distance of 3.3 pc (about 10 *l*-yr) for the zodiacal cloud, the Merope cloud, and the general interstellar dust cloud.

tion of the weakening of starlight leads to a discussion of its effect on the diffuse galactic light. Just as the interplanetary dust produces the zodiacal light by scattering sunlight, so dust-scattered starlight produces the diffuse galactic light described in Chapter 5. There is the difference that, in the first case, the observer has a definite position close to a single light source (the Sun) whereas in the second case he is surrounded by a large number of light sources (stars). In both cases the observer is embedded within the domain of the scattering material.

The diffuse-galactic-light observations have been interpreted in terms of two general parameters – the albedo of the particles and the angular dependence of the scattering function. For this latter, the scattering function, $S(\phi)$, is often referred to a parameter, g,

$$S(\phi)=\frac{1-g^2}{(1+g^2-2g\cos\phi)^{3/2}}. \tag{6.2}$$

In intuitive terms, g is the mean cosine of the scattering angle, ϕ. In Figure 6-14 we show graphical examples of g, including $g=0$, corresponding to isotropic scattering, $g=0.5$ and 0.6 strongly forward-scattering. For $g=+1$, the scattering is completely forward-scattering. Negative values of g refer to cases of predominant backscatter.

Two recent approaches to the analytical problem are based on the following assumptions.

(1) A dust sheet with a distribution similar to the stars in and near the galactic plane (van de Hulst and de Jong, 1969)

(2) Dust distributions in which discrete blobs occur, sometimes referred to as a raisin-pudding model (Mattila, 1970).

In both types of analysis, unique solutions are not possible, since a family of albedo, a, and g will usually satisfy the observations. However, when solutions for different regions are combined, the intersections have been interpreted as representative of the

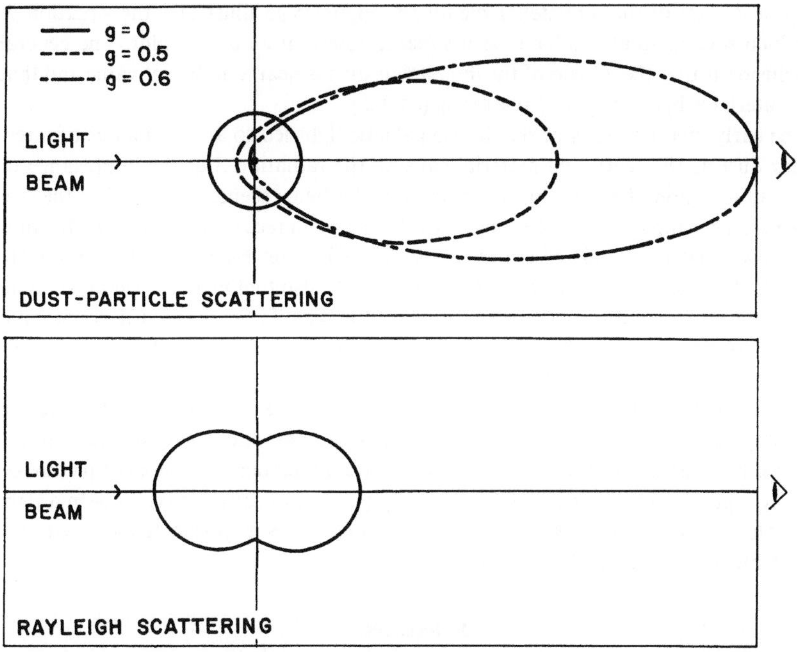

Fig. 6-14. Forward scattering of light by dust particles according to the parameter g (discussed in the text). In the lower part of the figure we repeat a part of Figure 1-2 to show the difference between scattering by dust particles and the very small (molecular) species responsible for Rayleigh scattering.

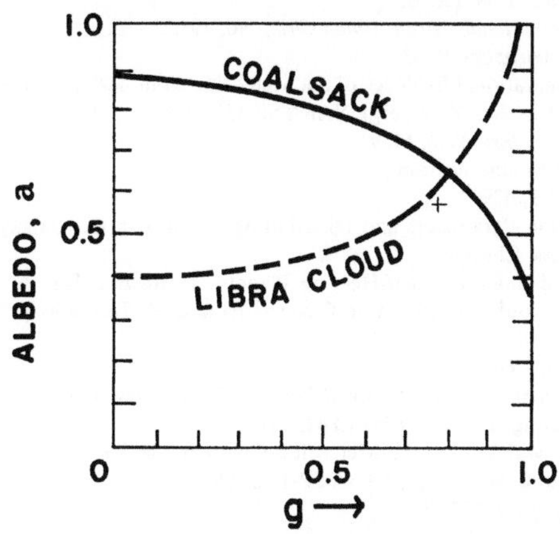

Fig. 6-15. Solutions of diffuse scattered light in the albedo (a) – forward-scattering (g) domain according to Mattila (1970). The intersection of the solutions for the Coalsack and the Libra cloud at $a=0.65$ and $g=0.8$ is in close agreement with a similar analysis of observations of two regions in the Milky Way (Taurus and Cygnus) marked by the + ($a=0.55$, $g=0.75$), based on observations by Witt (1968); see also Figure 5-4.

physical nature of the interstellar medium*. Figure 6-15 illustrates the situation in in which several solutions have been superimposed on an a vs g plot. The general conclusion is that the albedo of the interstellar dust is near a midpoint value and that the material is highly forward-scattering ($g > 0.5$).

The early investigations of the diffuse galactic light led to an acceptance of interstellar dust as the scattering material causing the radiation, but the importance of this manifestation of matter in the astronomical scheme seemed as trivial as the size of the grains compared to such grand objects as stars. In recent decades the realization has grown that both ends of the mass spectrum (Figures 6-2 and 6-3) are not only important in our snapshot views of the universe but also that in an astronomical time scale there is an evolution of dust/gas → star → dust/gas. The *average* turnaround time of this process as dusty material coalesces into stars and stars live out their times and return to dust has been estimated to be 2×10^9 yr. Fortunately, our star-Sun has a longevity sufficient to permit us a spectator look at these cosmic events. During the 5×10^9 yr of our planet 'we' have witnessed some two-and-a-half stardust turnarounds. In a particular case it may not be apparent whether our snapshot picture of a dusty region is due to dust that has recently (astronomically speaking) been a part of an old star or is coalescing into a new one. In some cases it appears that discrete and well-defined blobs are 'birthing' stars.

References

Bandermann, L. W.: 1967, *Physical Properties and Dynamics of Interplanetary Dust*, Tech. Rept. No. 771. U. Maryland, Dept. of Physics and Astronomy, College Park, Md.

Gillett, F. C.: 1966, *Zodiacal Light and Interplanetary Dust*, NASA Contr. NsG-281-62 and NONR 710(22), Rept. 2.

Kordylewski, K.: 1961, *Acta Astron.* **11**, 165.

Kordylewski, K.: 1966, *International Astron. Union Circ.*, No. 1985.

Mattila, K.: 1970, *Astron. Astrophys.* **9**, 53.

Millman, P. M.: 1967, 'Observational Evidence of the Meteoritic Complex', in *The Zodiacal Light and the Interplanetary Medium* (ed. by J. L. Weinberg), NASA SP-150, 399.

Oort, J. H.: 1950, *Bull. Astron. Inst. Neth.* **11**, 91.

Roach, J. R.: 1973, private communication.

Roosen, R. G.: 1968, *Icarus* **9**, 429.

Schmidt, M.: 1965, 'Rotation Parameters and Distribution of Mass in the Galaxy', Ch. 22 in *Galactic Structure*, U. Chicago Press, Chicago.

Shapley, H.: 1962, 'Crusted Stars and Self-Heating Planets'. in *Revista*, Ser. A, *Matematica y Fisica Teorica*, Vol. XIV, No. 1 and 2, Facultad de Ciencias Exactas & Technologia Tucuman, Republica Argentina, 69.

Simpson, J. W.: 1967, *Physics Today* **20**, 39.

van de Hulst, H. C.: 1957, *Light Scattering by Small Particles*, Wiley, New York.

van de Hulst, H. C. and de Jong, T.: 1969, *Physica* **41**, 151.

Whipple, F. L.: 1967, 'On Maintaining the Meteoritic Complex', in *The Zodiacal Light and the Interplanetary Medium* (ed. by J. L. Weinberg), NASA SP-150, 409.

Witt, Adolf N.: 1968, *Astrophys. J.* **152**, 59.

* Navigators may see a similarity with the method of determining a terrestrial position by the intersection of plots of two line-of-position determinations.

CHAPTER 7

COSMIC LIGHT AND COSMOLOGY

Every thinking person is a cosmologist. The child ponders space and time and asks profound questions about their beginning and end. The adult scans the night sky and propounds the same questions, answers to which often include philosophical excursions. The optical astronomer makes quantitative measurements such as the red shift of the spectral lines of distant, extragalactic objects and notes that the shift increases with increasing distance of the object. From this simple (in principle) but difficult (in practice) measurement, a whole generation has become familiar with the concept of an expanding universe.

The study of the light of the night sky is an exercise in exact photometry, but beyond that it is an adventure in space and time that challenges the most nimble imagination. By far the largest spatial and temporal leap is the one under scrutiny in this chapter, involving, as it does, an attempt to evaluate the light from the totality of objects external to our Galaxy, the cosmic light. The establishment of the fact that many of the 'nebular' appearing telescopic objects are actually external to our Galaxy (island universes) is one of the outstanding achievements of the 20th century.

The measurement of the cosmic light from our terrestrial platform is complicated – almost obviated – by the photometric dominance of what, in this chapter, is the 'foreground' light – the nightglow, the zodiacal light, and the integrated starlight and the associated diffuse galactic light from our own Galaxy*. Here it is helpful to summarize typical values for the several competing components of the light of the night sky in the direction toward the galactic pole – the most favorable for the discrimination of the cosmic light (see Figure 7-1).

Understandably, attempts at measuring the cosmic light from Earth have not been

LONS component	S_{10}(vis) units
Nightglow	50
Zodiacal light	117
Integrated starlight	
from $m = 5$ to ∞	28
$m = 15$ to ∞	2.6
$m = 20$ to ∞	0.15
Diffuse galactic light	9 (estimate)
Cosmic light	0.9 (de Vaucouleurs, 1949)

* From Earth's surface, the problem is further complicated by the difficulty of correcting for the extinction and scattering of our atmosphere, though this does not contribute a competing primary component.

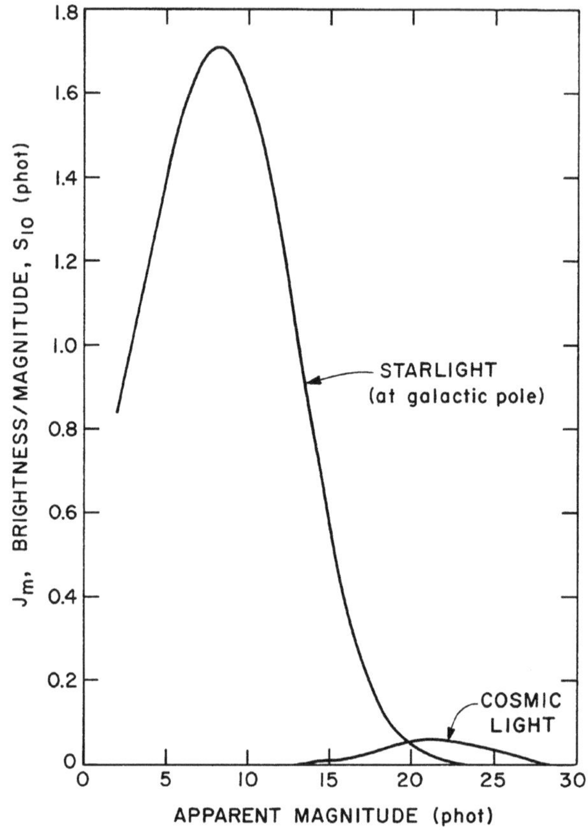

Fig. 7-1. Comparison of the light contribution from integrated starlight and from cosmic light as a function of apparent magnitude, in the direction of the galactic pole (adapted from de Vaucouleurs, 1949).

very successful. An empirical approach to the problem is to make counts of photographable nebulae to the observable limit and to extrapolate for the fainter objects. Such an estimate has been made by de Vaucouleurs (1949), who found that the surface brightness should be about 0.5 S_{10}(phot), corresponding approximately to 0.9 S_{10}(vis). Measurement of such a low level of brightness seems a difficult, if not impossible task. We recall the difficulty of measuring accurately the diffuse galactic light (Chapter 5) – the predicted brightness of the cosmic light is about 2% that of the diffuse galactic light. One attempt by Roach and Smith (1968) gave negative results – that is, a definitive value was not determined – but an upper limit of 5 S_{10}(vis) was reported (compare with the prediction of 0.9 by de Vaucouleurs).

There is possibly a more practical ploy that is being explored. The progressive red shift of the more and more distant extragalactic objects implies that the brightness of cosmic light relative to the foreground light increases as the observations are made at relatively long wavelengths. Thus, measurements in the IR could be fruitful.

It was the fond hope of many astronomers, during the planning decade before the start of the space age in 1958, that man's escape from Earth's surface would make possible an exact measurement of the cosmic light. In this chapter we shall explore this hope by way of a series of plots that will serve the double purpose of examining that possibility and at the same time taking us to the very horizon of the universe as presently known.

Any attempt to comprehend an overview of the great span of physical or temporal dimensions in our universe encounters the limitation of the human imagination, which is comfortable only with concepts in the general domain of human experience. The imagination deals satisfactorily with linear concepts – for example, plots in which both the x and y coordinates are directly proportional to the quantities under consideration. It is often impossible, however, to illustrate a particular domain by such a plot, and it is then frequently useful to make use of a logarithmic scale, in which successive integral steps on the plot correspond to a change by a factor, commonly ten.

Such a log plot is shown in Figure 7-2, in which we desire to take advantage of our recently acquired appreciation of the distance to the Moon by a plot showing how an astronaut en route to the Moon passes through the nightglow layer. On a linear plot drawn to scale, this would be unfeasible, since the Moon is some 60 R_E distant, and the nightglow is produced in an atmospheric 'skin' of a thickness of a few percent of an R_E. The nightglow-to-Moon span is conveniently shown in Figure 7-2, however, by the use of five logarithmic steps.

The unmanned satellites (such as OSO-6) have orbits at a height just above the nightglow at 400 km (Figure 7-2), but photometric observations from such a satellite might be affected by the nightglow, especially if the observation were made tangential to Earth. It is obvious, however, that current space technology makes possible observations from points sufficiently distant from Earth to be uncontaminated by the nightglow. Thus a first step has been taken in decontaminating our data toward the goal of evaluating the cosmic light. In one experiment alone (on OSO-6), several million observations of the 'night' sky have been made and are under intensive analysis.

We have placed two scales on Figure 7-2, giving the distance from Earth's surface in both kilometers and light-time (l-ms), as we wish to emphasize in this chapter the great expanse of our Universe not only in space but also in time. Note that the light-time from the portion of the upper atmosphere producing the nightglow is of the order of 1 ms and from the Moon about 1 s.

The excursions to the Moon in the manned space program, as well as the unmanned orbital satellites, have made possible the exclusion of the nightglow but have left in the data the zodiacal light, which, in large part, originates in a dust cloud extending from the Sun to envelop the Earth-Moon system and extend outward along the ecliptic plane. Figure 7-3 encompasses the part of the solar system in which the zodiacal cloud is concentrated. The zodiacal light, moving outward in the solar system, is assumed to

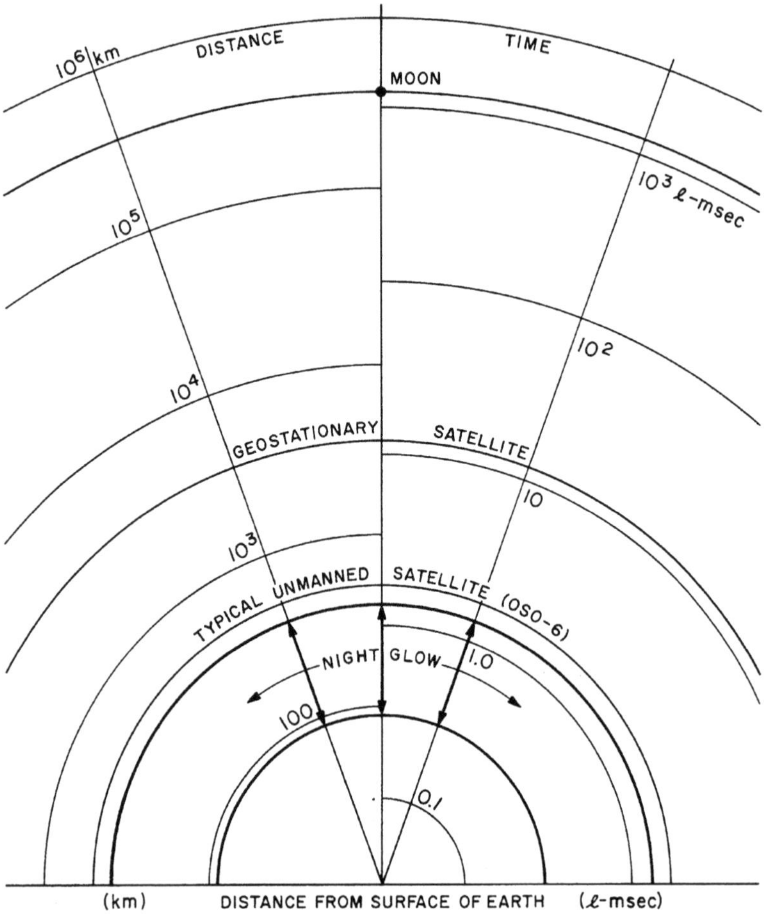

Fig. 7-2. Schematic representation of near space on a logarithmic scale. Distances are indicated on the left and light-time on the right.

decrease in brightness because of the probable decrease in the local density of dust as we proceed out past Mars toward Jupiter (see Figure 6-9), and also to the fact that the solar illumination on the dust decreases with distance from the Sun – at the distance of Jupiter (about 5 AU) the illumination is only about 1/25 that at Earth. Unmanned satellites presently en route or planned to the vicinity of Jupiter and beyond thus move into a region in which the zodiacal light is much less bright than it appears to the terrestrial observer. Thus a second step in the decontamination of the photometric data is within the technology of the 20th century.

To free the observations from the photometric contributions due to the starlight and the associated diffuse galactic light calls for such a gigantic leap that we are

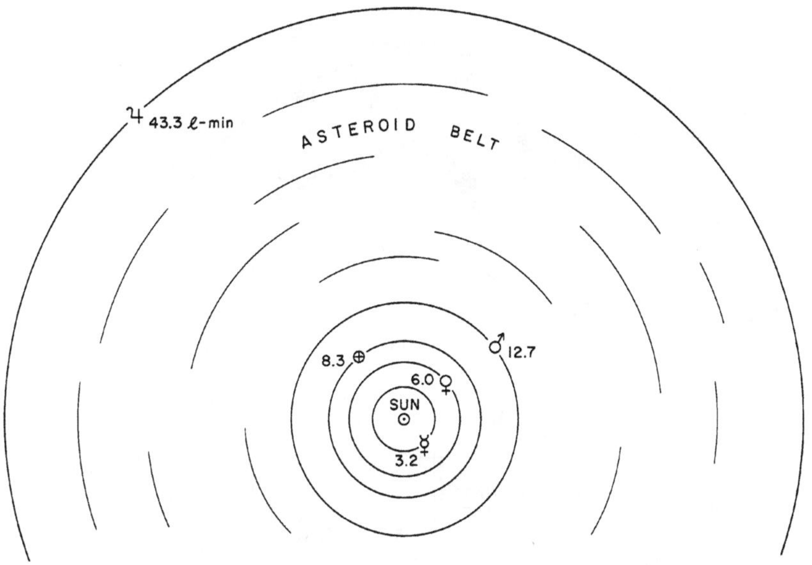

Fig. 7-3. Portion of solar system within the orbit of the planet Jupiter. Distances are shown in *l*-min.

showing it in three successive illustrations, each on a linear scale in itself but separated from each other by factor jumps in scale.

Figure 7-4 shows the distribution of known stars and star systems within 17 *l*-yr of the Sun. The solar system proper is a small dot on this scale – the orbit of Pluto (5.6 *l*-hr) being about 10^{-4} the distance to the nearest drawn circle (5 *l*-yr). The solar gravitational domain extends outward until a neighboring star's gravitation exceeds that of the Sun. This domain probably contains a myriad of objects that are too small to be self-luminous and not sufficiently numerous to result in absorption or scattering that would produce observable effects in the night sky (see Chapter 6). The stars plotted in Figure 7-4 include ten that are brighter than $m=6$ and thus are visible; these stars would remain individually visible to a traveler anywhere within the sphere of the plot. Further, the Sun would also become a visible star to add to the general sky – at a distance of 5 *l*-yr, the Sun would be a 0.7-mag. star (about as bright as Altair); at a distance of 10 *l*-yr, a 2.2-mag. star (Polaris); and at 15 *l*-yr, a 3.1-mag. star (a little brighter than ε Eridani, 3.7, and τ Ceti, 3.5). It is of passing interest that ε Eridani, at a distance of 10.7 *l*-yr, and τ Ceti (11.9 *l*-yr) are stars physically rather similar to the Sun and were prime candidates in a search for signals from possible intelligent civilizations in a program called, appropriately, OZMA.

Let us return to our purpose in showing Figure 7-4 – the search for a location at which the cosmic light might be isolated from the remaining foreground light from galactic stellar sources. As mentioned, the ten stars that are visible to terrestrial observers – plus the Sun – would remain as discrete stars. The light from the 34 fainter stars in the plot would total to the equivalent of 104 10th-mag. stars, which, if spread

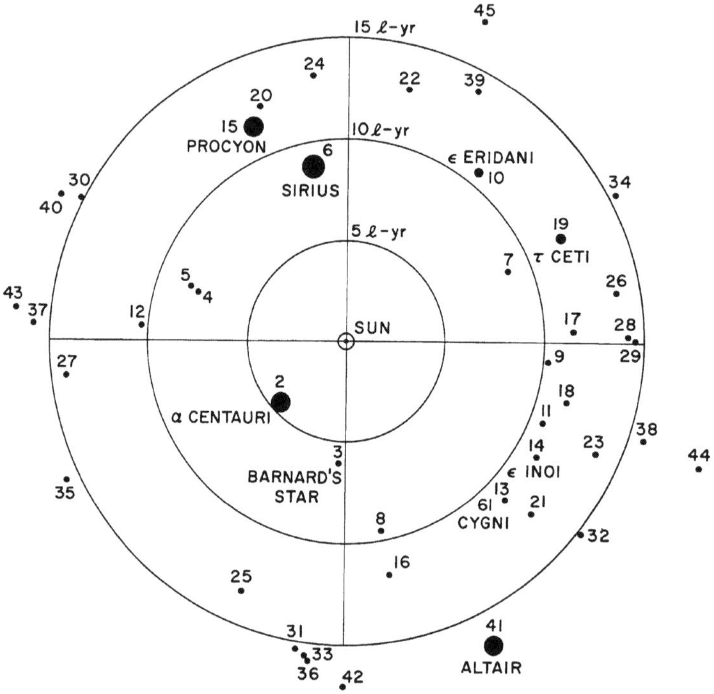

Fig. 7-4. Known stars within 17 *l*-yr of the Sun (in a sphere, projected on a plane). The numbers correspond to the order of radial distance from the Sun. The angular position on the plot is the star's right ascension. Stars that appear close together on the plot are not necessarily close in space, since the third dimension – for example, the star's declination – cannot be indicated (taken from van de Kamp, 1971 : 103, Table 1).

over the entire celestial sphere (41 253 sq deg), would contribute the equivalent of only 0.0025 10th-mag. star per sq deg – only a trivial fraction of the total starlight that concerns us in our cosmic search.

Thus an expedition to any spot in the local stellar domain would not be helpful – the Milky Way would still dominate the night sky. This conclusion is, of course, obvious with our present knowledge of our galactic system. After all, 15 *l*-yr is a trivial distance in terms of galactic dimensions, as shown in Figure 7-5, in which we note that the distance from the Sun to the galactic center is some 10 kpc (32 600 *l*-yr). With respect to the outer circle of Figure 7-5, with its radius of 15 kpc (49 000 *l*-yr), the outer circle of Figure 7-4 (15 *l*-yr) would reduce down to a circle smaller than the dot indicating the Sun.

It is obvious that the next step in our odyssey is to leave our Galaxy, which we do in Figure 7-6. Three of the objects shown in Figure 7-6 are visible – the two Magellanic clouds and the nebula in the constellation of Andromeda (Figure 7-7). The former are conspicuous, patch-like phenomena well-known to residents of the southern hemi-

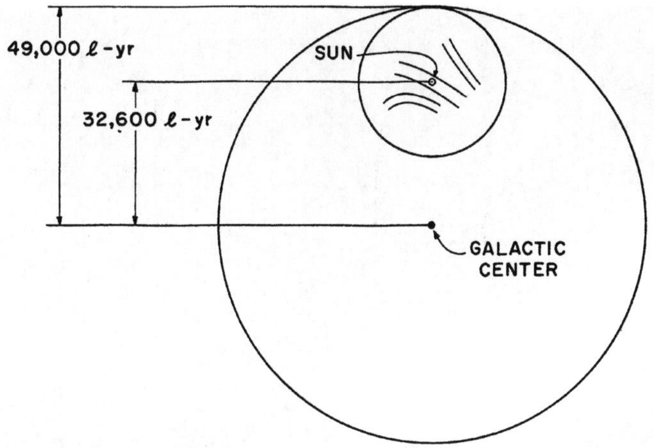

Fig. 7-5. Schematic representation of dimensions of our Galaxy.

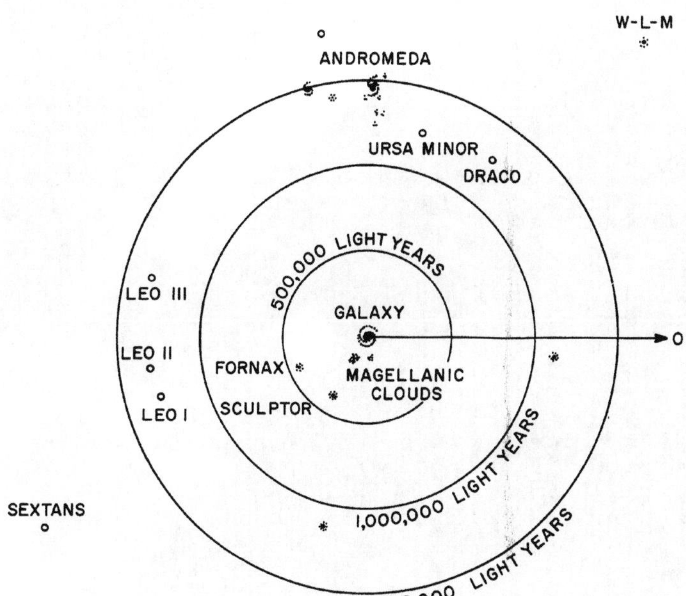

Fig. 7-6. Nearby extragalactic objects plotted with our Galaxy in the central position.

sphere. The latter appears as a 4th-mag. object; its 'nebular' appearance is easily seen with a small telescope, or even with binoculars. Clearly, an observer located at random in the domain covered by Figure 7-6 is in the ideal location for the measurement of cosmic light. For example, a point at the apex of an equilateral triangle (libration

Fig. 7-7. The extragalactic nebula in Andromeda.

point) equidistant from the Andromeda nebula and our Galaxy would reveal each of these two objects as of the 4th mag., and the rest of the sky would be only as bright as the totality of the light from the other and more distant galaxies.

Here we might interrupt to recall the dilemma posed by Olbers (see Chapter 2), with extragalactic objects replacing 'suns' in the argument. We may paraphrase Olbers (1826):

If there really be extragalactic nebulae in the whole of space, and to infinity, and if they are placed at equal distances from each other, or grouped into systems, their number must be infinite, and the whole vault of heaven should appear as bright as a typical extragalactic object; for every line which may be supposed to emanate from our eye towards the sky, would necessarily meet an extragalactic object, and thus every point of the sky would bring to us a ray of cosmic light.

Under these assumptions, the cosmic light should present us with a surface brightness equal to that of a typical object such as the Andromeda nebula. But, according to the integration of de Vaucouleurs, the cosmic light is some 40 times fainter. The full significance of this fact is confused by the systematic red shift of the light of the extragalactic objects.

Before making our final leap, note the large jump in the scales between Figures 7-5 and 7-6 – from the 49 000 l-yr for the larger circle in Figure 7-5 we make a more-than-10-fold jump in going to the smallest circle of Figure 7-6. As in all of our transitional steps, the earlier domain is dwarfed by the succeeding one.

A rough idea of the nature of the problem of the cosmic light may be appreciated by a study of Figure 7-8, in which we again change the scale by a large factor, this time

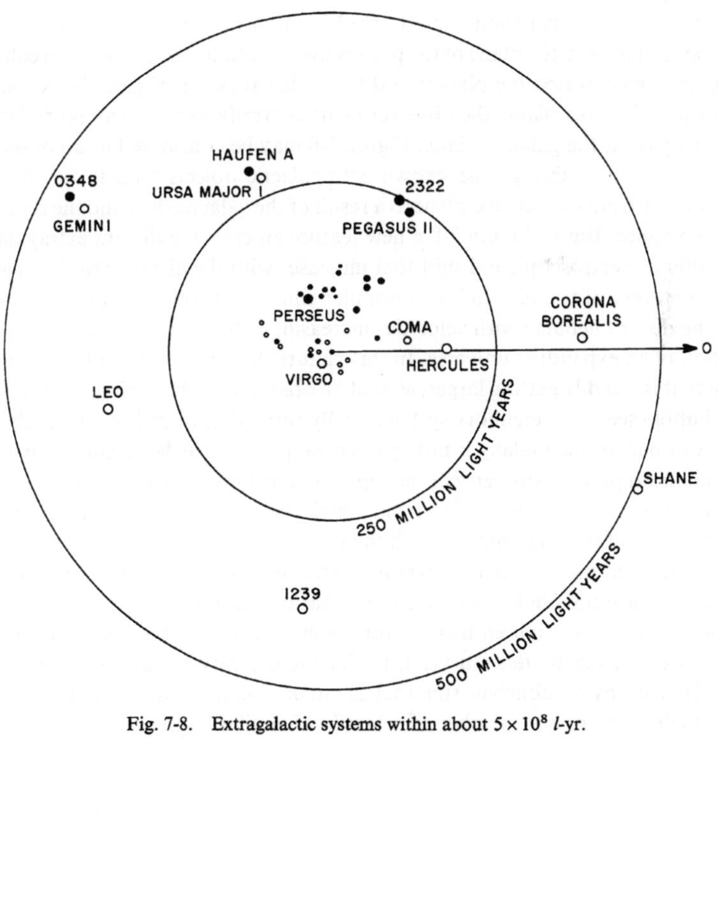

Fig. 7-8. Extragalactic systems within about 5×10^8 l-yr.

by about 300, stretching the imagination as we contemplate the fact that the light from many of the plotted objects originated at a time when our Galaxy was about two revolutions back (period of revolution of our Galaxy is about 2×10^8 yr), and millions of years before the appearance of even primitive astronomers on our planet. By no means are all of the extragalactic objects within a sphere of 5×10^8 yr shown in Figure 7-8. For example, there are severe observational limitations in obtaining a complete count – the circle of the Milky Way of our Galaxy constitutes a 'zone of avoidance' as far as extragalactic objects are concerned because of the interposition of the galactic dust concentrated in and near that plane. Observation of the extragalactic domain becomes more and more effective as the observer looks progressively away from the Milky Way and towards the galactic pole – here is where an accurate photometric measurement of the cosmic light could be helpful to the cosmologist.

We have called attention to the drastic scalar jumps made as we moved from the Earth-Moon system (Figure 7-2) to the solar system (Figure 7-3), to the nearby-star domain (Figure 7-4), and to galactic associations (Figure 7-8). Another figure that might have been added would have had as its outer arc the horizon of our knowledge of the universe – say 10^{10} *l*-yr – another scale change of 20 over that of Figure 7-8.

It should be mentioned that not all the figures have the same physical significance. Figure 7-2 involves the observer in the center of a sphere looking outward through the nightglow layer in logarithmic steps to the Moon's distance. Figure 7-3 is basically a two-dimensional plot, the plane of the page being essentially the plane of the ecliptic, in which are concentrated the planets and the zodiacal cloud. Figure 7-4 is again a sphere projected onto a plane, the observer being centrally located. In Figure 7-5 the plane of the page is the galactic plane. Figure 7-6 may be considered as an observer-centered sphere, even though the known extragalactic objects tend to be *observed* selectively away from the galactic plane as a result of the galactic dust and the resulting zone of avoidance. But in Figure 7-8 a new feature enters – the distant extragalactic objects suffer a spectroscopic red shift that increases with the distance to the objects. The interpretation of the red shift as a doppler shift – implying the outward movement of the distant nebulae with velocities increasing with distance – has dramatized the concept of an expanding universe. In this picture the observer is on the surface of a bubble that is steadily getting larger, so that an observer on the surface of this hypothetical bubble sees his neighbors systematically retreating from him. Thus the observer in our part of the Galaxy is not in a central position, as he might egotistically think from the apparent symmetry of the repulsive motions away from him!

We have now explored the region from which the cosmic light originates via the dramatic vehicle of the imagination, with the subterfuge that we might find a spot for making the desired measurement. Several illustrations back we had actually far exceeded the physical possibility of such an observational mission, except under the supervision of a science-fiction enthusiast, for whom 'the impossible takes only a little longer'. So we practical writers must return, albeit reluctantly, to our small blue planet (number 3) ruled by a suburban star that is circulating around the inner city of a Galaxy which, in turn, is a member of a family of similar aggregations.

As we have seen, the elimination of the light from the nightglow has already been accomplished aboard both unmanned and manned satellites during the first 14 yr of the space age. Furthermore, present satellite programs involving expeditions to and beyond Jupiter promise the reduction of the zodiacal light to low levels of brightness. Thus it is likely that, by the end of the first few decades of the space age, measurements of the cosmic light will become available. The reduction of the integrated starlight suggested by the three entries in our earlier table is illustrated in Figure 7-1, in which the contribution from integrated starlight to the night-sky light is shown contrasted with that from cosmic light based on de Vaucouleurs' integration. It is obvious that a large potential gain is made available by going to fainter stars for the start of the integration. This approach calls for the use of a large telescope, for which faint stars appear individually rather than as part of the background, leaving the diffuse galactic light as the final photometric contaminant to an observer in the galactic system.

The prognosis for a successful assault on the cosmic-light measurement problem is thus clear – observe toward the galactic pole with a large telescope on a satellite in the outer part of the solar system – hint to astronomers who may now inhabit one of the satellites of Jupiter or Saturn!

In spite of the difficulty in disentangling the several components of the light of the night sky, and especially of extracting the cosmic light, we wish to emphasize our felicitous astronomical location in that no individual component of that light predominates so completely that all of the others are lost in the glare of the one. It is possible to imagine circumstances such that the sky would hide Nature's secrets rather than reveal them. For example, the zodiacal cloud may have been much more dense during early periods of our solar system (and of course might become more dense in the future), so dense that the zodiacal light would be so bright as to restrict or even preclude sidereal, galactic, and extragalactic research. The Milky Way, under such circumstances, would be so overwhelmed photometrically as to take away from us one of the aesthetically beautiful features of our planetary life. The same Milky Way loss would have occurred if the OH nightglow had been concentrated in the visual part of the spectrum rather than in the IR, in which the human eye is insensitive. Without the visibility of the Milky Way, would William Herschel have been inspired to gage the stars, or J. C. Kapteyn to set up the system of systematic star counts that led to our understanding of galactic structure?

These comments are made to temper our disappointment that we have to admit that the integrated cosmic light has been inadequately measured. We have only recently partly escaped physically from our terrestrial prison into nearby space and have tried in this book a more adventurous foray by way of outward glimpses through our atmospheric prison walls.

References

de Vaucouleurs, G.: 1949, *Ann. Astrophys.* **12**, 162.
Olbers, W.: 1826, *Edinburgh New Phil. J.* **1**, 141.
Roach, F. E. and Smith, L. L.: 1968, *Geophys. J., Royal Astronomical Soc.* **15**, 227.
van de Kamp, P.: 1971, *Ann. Rev. Astron. Astrophys.* **9**, 103.

EPILOGUE

Finally, we are impressed with the fact that a glance at the night sky is an adventure in both space and time. The cosmos is not so much at our fingertips as at our eyeballs – the multiple aspects of the universe present themselves freely to us every night. The incredibly varied sources of energy in the universe emit light quanta that impinge on our tiny retinal nerve endings after traversing both spaces and time periods so vast that we can visualize them, in a crude fashion, only with the help of numbers that are traumatically large.

A single visual sweep of the night sky brings us all sorts of light – light that left the nightglow zone, a mere 100 km distant, only about 1 ms ago (Figure E-1). Light that left our nearest neighboring star, α Centauri, about 4 yr ago; light that left the brilliant

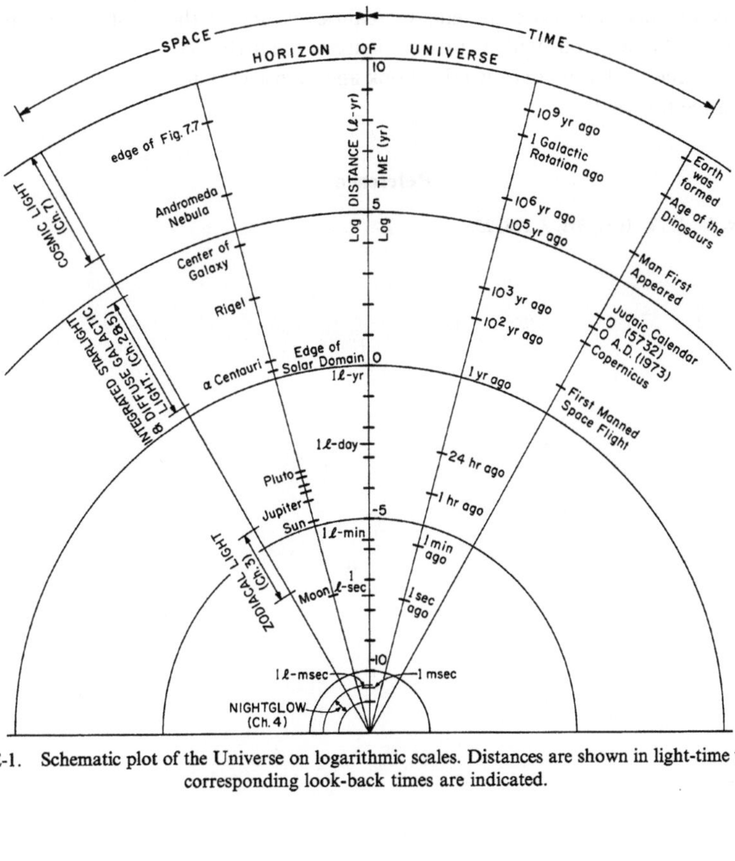

Fig. E-1. Schematic plot of the Universe on logarithmic scales. Distances are shown in light-time units; corresponding look-back times are indicated.

star Rigel about the time of Copernicus. Light that left one of our neighboring galaxies, the beautiful Andromeda nebula, about the time Man first appeared on Earth. And light that left some of the most distant nebulae – say 3000 Mpc (10^{10} *l*-yr) distant, at about the time Earth was formed.

The reader interested in the possibility that astronomers may now be seeing 'to the edge of the world' in both space and time might profitably read Sandage (1972). In that paper (and earlier ones in the series) there is a critical discussion of the relationship between the observed spectroscopic red shift and the distance-time of the object under scrutiny.

The light we see represents its source at the time the light started on its journey. What has happened to that source in the time since? What has happened to the Andromeda nebula in the million years since the light we see tonight left there? And how about Rigel, whose light left only 500 yr ago – a fleeting moment in astronomical time?

We have the whole of past time before our wondering eyes at night – and the old and the young quanta arrive together to give an illusion of simultaneity. Sorting out the illusions and fastening onto the elusive realities is one of the most fascinating of puzzles – and it may start with a cursory glance at the night sky.

We urge the reader to put down this book and take such a glance – on a clear night, you can see forever!

Reference

Sandage, A.: 1972, *Astrophys. J.* **178**, 25.

SUBJECT INDEX

GEOPHYSICS AND ASTROPHYSICS MONOGRAPHS

AN INTERNATIONAL SERIES OF FUNDAMENTAL TEXTBOOKS

1. R. Grant Athay, *Radiation Transport in Spectral Lines*. 1972, XIII + 263 pp.
2. J. Coulomb, *Sea Floor Spreading and Continental Drift*. 1972, X + 184 pp.
3. G. T. Csanady, *Turbulent Diffusion in the Environment*. 1973, XII + 248 pp.

Forthcoming:

5. R. Grant Athay, *The Solar Chromosphere and Corona*
6. J. Iribarne and W. Godson, *Atmospheric Thermodynamics*
7. Z. Kopal, *The Moon in the Post-Apollo Era*
8. Z. Švestka and L. De Feiter, *Solar High Energy Photon and Particle Emission*
9. A. Vallance Jones, *The Aurora*
10. R. Newell, *Global Distribution of Atmospheric Constituents*
11. G. Haerendel, *Magnetospheric Processes*

MIX
Papier aus verantwortungsvollen Quellen
Paper from responsible sources
FSC® C105338

If you have any concerns about our products,
you can contact us on
ProductSafety@springernature.com

In case Publisher is established outside the EU,
the EU authorized representative is:
Springer Nature Customer Service Center GmbH
Europaplatz 3, 69115 Heidelberg, Germany

Printed by Libri Plureos GmbH
in Hamburg, Germany